B

Progress in Computer Science
No. 5

Edited by
J. Bentley
E. Coffman
R. L. Graham
D. Kuck
N. Pippenger

Birkhäuser
Boston · Basel · Stuttgart

Nachum Dershowitz
The Evolution of Programs

1983

Birkhäuser
Boston • Basel • Stuttgart

Author:

Nachum Dershowitz
Department of Computer Science
University of Illinois
at Urbana-Champaign
1304 West Springfield Avenue
Urbana, Illinois 61801-2987, USA

Library of Congress Cataloging in Publication Data

Dershowitz, Nachum.
    The evolution of programs.

    (Progress in computer science ; no. 5)
    Bibliography: p.
    Includes index.
    1. Electronic digital computers—Programming.
I. Title.   II. Series.
QA76.6.D465   1983   001.64'2   83-15770
ISBN 0-8176-3156-9
ISBN 0-8176-3171-2 (pkb.)

CIP-Kurztitelaufnahme der Deutschen Bibliothek

Dershowitz, Nachum:
The evolution of programs / Nachum Dershowitz.
- Boston ; Basel ; Stuttgart : Birkhäuser, 1983.

    (Progress in computer science ; Vol. 5)
    ISBN 3-7643-3171-2 (Basel, Stuttgart) brosch.;
    ISBN 0-8176-3171-2 (Boston) brosch.;
    ISBN 3-7643-3156-9 (Basel, Stuttgart) Pp.;
    ISBN 0-8176-3156-9 (Boston) Pp.

NE: GT

© Birkhäuser Boston, Inc., 1983
ISBN 0-8176-3156-9 (hardcover); 0-8176-3171-2 (paperback)
ISBN 3-7643-3156-9 (hardcover); 3-7643-3171-2 (paperback)
Printed in USA

9 8 7 6 5 4 3 2 1

In memory.

Preface

עשׂוֹת סְפָרִים אֵין קֵץ, וְלַהַג הַרְבֵּה יְגִעַת בָּשָׂר.

—Ecclesiastes 12:12

Programs are invariably subjected to many forms of transformation. After an initial version of a program has been designed and developed, it undergoes debugging and certification. In addition, most long-lived programs have a life-cycle that includes modifications to meet amended specifications and extensions for expanded capabilities. Such evolutionary aspects of programming are the topic of this monograph. We present formal methods for manipulating programs and illustrate their application with numerous examples. Such methods could be incorporated in semi-automated programming environments, where they would serve to ease the burden on the programmer.

We begin by describing a method whereby a given program that achieves one goal can be modified to achieve a different goal or a program that computes wrong results can be debugged to achieve the intended results. The abstraction of a set of cognate programs to obtain a program schema, and the instantiation of abstract schemata to solve concrete problems, are approached from the same perspective.

In addition, we describe synthesis rules for generating code from specifications and annotation rules for making assertions about code. The synthesis rules may be used when a program is first being developed, or when, in the course of modifying a program, the need arises to rewrite a program segment. Annotation rules may be used for the purpose of determining what an incorrect program really does before attempting to debug it or how a correct program works before attempting to modify it.

The first chapter introduces the topics. It is followed by a general overview of the various aspects of program manipulation; their individual roles and close interaction are illustrated in an account of the evolution of one example program. The remainder of this book is composed of chapters on

- modification and debugging
- abstraction and instantiation
- synthesis and extension, and
- annotation and analysis.

Each chapter is largely self-contained, with a common set of examples threaded through them. The reader is presumed only to know how to program and to have some facility with predicate logic. The concluding chapter discusses the role of these methods in the broader context of programming environments. Bibliographic remarks are included in each chapter.

This monograph is based on my Ph.D. dissertation (submitted in October 1978 to the Department of Applied Mathematics and Scientific Council of the Weizmann Institute of Science) and on four published papers: modification and debugging are discussed in [Dershowitz-Manna77]; abstraction and instantiation are the topic of [Dershowitz81]; synthesis is discussed in [DershowitzManna75]; annotation is the subject of [DershowitzManna81].

Acknowledgments

There are many to whom I wish to take this opportunity to express my deep gratitude.

I sincerely thank Zohar Manna, my thesis advisor, and Richard Waldinger for their constant encouragement and crucial guidance.

I thank the Weizmann Institute of Science, the Artificial Intelligence Laboratory of Stanford University, SRI International, and the University of Illinois for the comfortable working environments they provided.

For their support during various stages of this work, I thank the United States Air Force Office of Scientific Research (Grant AFOSR-76-2909), the National Science Foundation (Grants MCS-76-83655 and MCS-79-08497), and the Advanced Research Projects Agency of the Department of Defense (Contract MDA 903-76-C-0206).

To those who took the time to read and comment on various versions of parts of this work, I am most grateful. They include: John Darlington, Steven Greenbaum, David Harel, Alan Josephson, Shmuel Katz, Jim King, Fei-Pei Lai, Yuh-Jeng Lee, Arvin Levine, Larry Paulson, Wolfgang Polak, Yehoshua Sagiv, Adi Shamir, and anonymous reviewers.

I thank my parents for so much.

And thank you, Schulamith, Erga, and Idan, for those other things in life.

Table of Contents

8

Table of Figures

Table of Programs

Chapter 1

Introduction

Ex nihilo fit nihil.

—Epicurus

Programming begins with a specification of what the envisioned program ought to do. It is the programmer's job to develop an executable program satisfying that specification. Yet, only a small fraction of a programmer's time is typically devoted to the creation of original programs *ex nihilo*. Instead, most of his effort is normally devoted to such activities as debugging incorrect programs, adapting known techniques to specific problems at hand, modifying existing programs to meet amended specifications, extending old programs for expanded capabilities, and abstracting ideas of general applicability into "subroutines."

The goal of research in "automated programming" is to formalize methods and strategies used by programmers in going about their work so that those ideas may be incorporated in automatic, or interactive, programming environments. In our view, program development systems should incorporate formal tools, not only for constructing programs from scratch, but also for transforming and manipulating programs. In this work, we illustrate how some of the *evolutionary* aspects of programming might be emulated by such a system. Chapter 2 presents an overview of our general approach.

Programmers improve with experience by assimilating techniques as they are encountered, and judiciously applying the ideas learned to new problems. One way in which prior knowledge can be used is by modifying a known program to achieve some new goal. For example, a program that uses the "binary-search" technique to compute square-roots might be transformed into one that divides two numbers. We show how to modify programs by first finding an analogy between the specification of the existing program and that of the program we wish to construct. This analogy is then used as a basis for transforming the existing program to meet the new specification. Program debugging is treated as a special case of modification: if a program computes wrong results, it must be modified to achieve the intended results. Modification and debugging are the subject of Chapter 3.

Program modification is not the only manner in which a programmer utilizes previously acquired knowledge. After coming up with several modifications of his first "wheel," he is likely to formulate for himself (and perhaps for others) an abstract notion of the underlying principle and reuse it in new, but related, applications. Program "schemata" are a convenient form for "remembering" programming strategies such as the "generate-and-test" paradigm or the "binary-search" technique. The specifications of a schema can be stated in terms of abstract entities that must satisfy certain conditions for the schema to apply in any given instance.

The abstraction of a set of concrete programs to obtain a program schema and the instantiation of abstract schemata to solve concrete problems may be viewed from the perspective of modification methods. This perspective provides a methodology for applying old knowledge to new problems. Beginning with a set of programs sharing some basic strategy, and their correctness proofs, a program schema that represents their analogous elements is derived. The resultant schema's abstract specification may be compared with a given concrete specification to suggest an instantiation that will yield a concrete program when applied to the schema. Not all instantiations, however, result in a correct program.

It is necessary to formulate preconditions under which the schema is applicable; only when an instantiation satisfies the preconditions, is the correctness of the new program guaranteed. Chapter 4 covers abstraction and instantiation.

Extending a program to satisfy additional specifications is another form of program modification. One must construct code that extends the incomplete program to achieve the new goals, while ensuring that the original specification continues to be satisfied. Modifications based on analogy and extension can be combined to solve a particular problem. The analogy between a new problem and a given program may suggest how to achieve part of the specified goal; the transformed program is then extended to achieve the remainder.

Of course, it is not always possible to find a program or schema that can be adapted to the problem at hand. And, even when one can modify an old program or instantiate a known schema, it may turn out that some program segments need to be reconstructed. In such cases, one has a specification for a desired segment, and the goal is to transform it step-by-step into executable code. When proceeding in a top-down fashion—as suggested by "structured programming" methodology—each step consists of rewriting a segment of the developing program in increased detail. If every step is transparent enough to ensure correctness, each partial program in the series is sure to be equivalent to its predecessor. In particular, the final program is guaranteed to satisfy the initial specifications. Synthesis and extension are discussed in Chapter 5.

To debug an incorrect program, it helps to know what the program really does, not just what it was intended to do. Moreover, various facts about a program are frequently needed to guide other forms of modification. For these purposes, it is convenient if the program is annotated with "assertions." Assertions are a useful means of documenting facts about the internal workings of a program; they relate to specific points in the program and assert that some relation holds for the current values of the program variables whenever control passes through that point. Given a program along with its input-output specification, we endeavor to annotate the program with assertions that explain the actual workings of the program, regardless of whether the program is correct.

These annotations can be used for guiding transformations, verifying correctness, or analyzing efficiency. Annotation and analysis are the topic of Chapter 6.

Ideally, we envision a semi-automated environment in which these types of programming activities are supported. The methods of program manipulation that we describe are for the most part amenable to such automation, and we have implemented many of them in an experimental, "proving-ground" system. These issues are briefly discussed in the last chapter.

Our QLISP implementation consists of three units: modifier, annotator, and synthesizer. The modifier has, for example, modified a square-root program to compute quotients and has debugged an incorrect division program. The annotator generated the necessary invariants for these examples, and for more complex programs, including one for sorting. The synthesizer successfully constructed several complete programs, such as one for finding the minimal element of an array. Some details of the implementation are included in Appendix 5. Other appendices contain compilations of transformations, schemata, synthesis rules, and annotation rules.

Chapter 2

General Overview

Analogy is a sort of similarity. Similar objects agree with each other in some respects, analogous objects agree in certain relations of their respective parts.

Inference by analogy appears to be the most common kind of conclusion, and it is possibly the most essential kind. It yields more or less plausible conjectures which may or may not be confirmed by experience and stricter reasoning.

Analogy pervades all our thinking, our everyday speech and our trivial conclusions as well as artistic ways of expression and the highest scientific achievements. Analogy is used on very different levels. People often use vague, ambiguous, incomplete, or incompletely clarified analogies, but analogy may reach the level of mathematical precision. All sorts of analogy may play a role in the discovery of a solution and so we should not neglect any sort.

—George Pólya (How to solve it)

2.1. Introduction

In this overview we trace the life-cycle of a single example program. The example—motivated by [Wensley59] and [Dijkstra76]—is outlined in Figure 1. This example illustrates some of the kinds of transformations programs undergo and how various aspects of program manipulation are interrelated. This chapter is meant to impart the overall flavor of our approach; detailed treatments of each aspect are left to the chapters that follow.

We begin with an imperfect program to compute the quotient of two real numbers. The program is *debugged,* after determining enough about what the program actually does by *annotating* it with assertions. Once the division program is corrected, it is *modified* to approximate the square-root of a real number. Underlying both the division and square-root programs is the binary-search technique; by *abstracting* these two programs, a binary-search schema is obtained. This schema is then *instantiated* to obtain a third program, one to compute the square-root of an integer. Part of that program is *synthesized* from scratch.

2.2. The Problem

Consider the problem of computing the quotient q of two nonnegative real numbers c and d within a specified (positive) tolerance e. These requirements are conveniently expressed in a high-level *assertion language* in two parts: an output specification and an input specification. The *output specification* states the desired relationship among the program variables upon termination. In our case, the output specification

$$|c/d-q|<e$$

indicates that the (absolute value of the) difference between the exact quotient c/d and the result q should be less than e. The *input specification* defines the set of inputs on which the program is intended to operate. Assuming that we wish our program to handle the case when

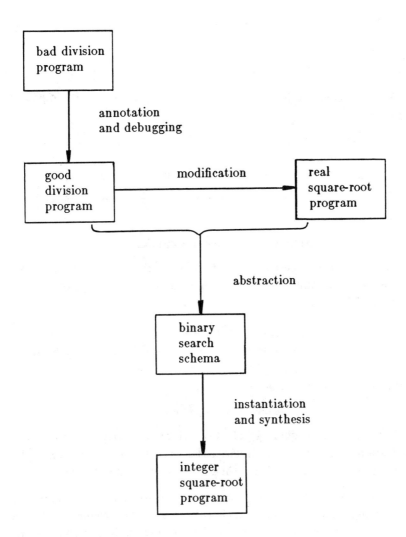

Figure 1. *Evolution of a division program.*

the quotient is in the range 0 to 1, an appropriate input specification might be[1]

$$0 \leq c < d \wedge e > 0.$$

We can express this programming goal in the form of the following skeleton program:

P_1: **begin comment** *real–division specification*
    assert $0 \leq c < d,\ e > 0$
    achieve $|c/d - q| < e$ **varying** q
    end

The **achieve** statement,

$$\textbf{achieve } |c/d - q| < e \textbf{ varying } q,$$

contains the output specification which gives the relation between the variables q, c, d, and e that we wish to attain at the end of program execution. The clause

$$\textbf{varying } q$$

indicates that of the variables in the specification, only q may be set by the program; the other variables, c, d, and e, contain input values that remain fixed. The **assert** statement,

$$\textbf{assert } 0 \leq c < d,\ e > 0,$$

attached to the beginning of the program, contains the input specification which gives the relation between the input variables that may be assumed to always hold when program execution begins.

An **achieve** statement may be considered to be a "very high-level" programming construct that "somehow" achieves the specified relation at that point in the program. It is not directly executable; the task of the programmer—be he human or machine—is to systematically transform the **achieve** statement into an executable program by replacing it with

---

[1]We will be using a predicate calculus notation in which the symbols $\sim$, $\wedge$, $\vee$, $\equiv$, $\supset$, $\forall$, and $\exists$ mean "not," "and," "or," "if and only if," "implies," "for all," and "there exists," respectively.

more detailed code. If the replacement itself contains **achieve** statements, then the process iterates, step by step, until a machine-executable program is obtained that contains only "primitive" statements and operators. In this manner, we use **achieve** statements to provide a formalism for the development of programs by "stepwise refinement" ([Wirth71]). The final program will be of the form

P_1: **begin comment** *real-division program skeleton*
    assert $0 \le c < d,\ e > 0$
    purpose $|c/d-q| < e$
    . . .
    code
    . . .
    suggest $|c/d-q| < e$
    end

The **purpose** statement,

$$\textbf{purpose } |c/d-q| < e,$$

is a comment describing the intent of the (indented) code following it. The statement

$$\textbf{suggest } |c/d-q| < e$$

contains the programmer's contention that the preceding code actually achieves the desired relation, i.e. the relation $|c/d-q| < e$ holds for the value of q when control reaches the end of the program.

For the problem at hand, we assume that no general real-division operator / is available, though division by powers of two ("shifts") is permissible. Otherwise, the problem could be solved with a trivial assignment statement

$$q := c/d.$$

Similarly, were it not for the restriction that only the variable q may be set by the program, the problem could be solved, for example, by setting both q and c to 0. This would satisfy the specification $|c/d-q| < e$, but is not the intended solution.

Now let us imagine that a programmer went ahead and constructed the following program:

T_1: **begin comment** *suggested real–division program*
  B_1: **assert** $0 \leq c < d,\ e > 0$
  purpose $|c/d - q| < e$
    purpose $q \leq c/d,\ c/d < q + s,\ s \leq e$
    $(q,s) := (0,1)$
      loop L_1: **suggest** $q \leq c/d,\ c/d < q + s$
      until $s \leq e$
        purpose $q \leq c/d,\ c/d < q + s,\ 0 < s < s_{L_1}$
        if $d \cdot (q + s) \leq c$ **then** $q := q + s$ **fi**
        $s := s/2$
      repeat
    suggest $q \leq c/d,\ c/d < q + s,\ s \leq e$
  E_1: **suggest** $|c/d - q| < e$
  end

The comment

$$\textbf{purpose } q \leq c/d,\ c/d < q + s,\ s \leq e$$

indicates that the programmer's intention is to achieve the desired relation $|c/d - q| < e$ by achieving the three subgoals $q \leq c/d$, $c/d \leq q + s$, and $s \leq e$. Achieving these relations is sufficient for $|c/d - q| < e$ to hold. For this purpose the programmer constructed an iterative loop intended to keep the first two relations invariantly true while making progress towards the third. The relations that are to remain true are contained in the statement

$$\textbf{suggest } q \leq c/d,\ c/d < q + s$$

at label L_1.[2] To make progress towards satisfying $s \leq e$, the goal of the loop body is

$$\textbf{purpose } q \leq c/d,\ c/d < q + s,\ 0 < s < s_{L_1},$$

where s_{L_1} denotes the value of the variable s when control was last at the label L_1. This means that the value of s is to be less than its value s_{L_1} at the head of the loop, while, at the same time, the relations

---

[2]The predicates in **purpose** statements correspond to those used in the "subgoal induction" method of program verification ([Manna71] and [MorrisWegbreit77]); the predicates in **assert** and **suggest** statements correspond to those used in the "inductive

$q \leq c/d$ and $c/d < q+s$ are to remain true. To accomplish that purpose, the two loop-body statements

$$\textbf{if } d \cdot (q+s) \leq c \textbf{ then } q := q+s \textbf{ fi}$$
$$s := s/2$$

are repeated (zero or more times) until the test

$$\textbf{until } s \leq e$$

becomes true, at which point the loop is exited.[3]

When a relation, such as $q \leq c/d$, has been *proved* to hold each time control passes through some point, then it is said to be an *invariant assertion* (or just *invariant*) at that point. As long as it has not been proved to hold, it is called a *candidate*. For the loop candidates

$$\textbf{suggest } q \leq c/d, \ c/d < q+s$$

to be loop invariants, they must hold when the loop is first entered and must remain true each subsequent time control returns to the beginning of the loop. They are established initially by the multiple assignment

$$(q,s) := (0,1)$$

of 0 to q and 1 to s, since both $0 \leq c/d$ and $c/d < 0+1$ are implied by the assumption $0 \leq c < d$. Unfortunately, though the two candidates hold when the loop is first entered, the candidate $c/d < q+s$ is not preserved when the loop is iterated; thus, it is *not* an invariant.

A relation, such as $|c/d - q| < e$, associated with the end of a program, is called an *output candidate*. For it to be an invariant at that point, the final values of the variables must always satisfy the suggested relation when the program terminates; it is termed an *output invariant* when this is the case. Consequently, a program is *correct*, if there exist output invariants that imply the output specification. In our case, the output candidate $|c/d - q| < e$ cannot be shown to be invariant. In fact, by running the program with $c=1$, $d=3$, and $e=1/3$, for instance, the

---

assertion" method of program verification ([Floyd67] and [Hoare69]).

[3]The **loop-until-repeat** construct we use is based on the suggestion of Ole-Johan Dahl in [Knuth74]; **achieve** statements were used in [Sussman75]. As a matter of convenience, we omit type declarations from our programs.

programmer would discover that the result $q=0$ does not satisfy $|c/d-q| < e$.[4] Since these values for the variables satisfy the input specification, but do not satisfy the output specification, the program, as given, is *incorrect*.

2.3. Annotation

We know what the above program was intended for, and we know that it does not always fulfill those intentions. However, before we can debug it, we need to know more about what it actually does. This will be accomplished by examining the code, trying to extract as many invariant relations between the variables as possible, and annotating the program with those invariants. Our methods of program annotation are discussed more fully in Chapter 6. There they are formulated as inference rules: the antecedents of each rule are annotated program segments and the consequent is either an invariant or a candidate. These rules have been implemented; for more details, see Appendix 5.

As a first step, note that the input variables c, d, and e are not changed by the program. Therefore the input assertion

$$B_1: \textbf{assert } 0 \leq c < d, \, e > 0$$

holds throughout execution of the program. Such an assertion is termed a *global invariant* of the program; we write

$$\textbf{assert } 0 \leq c < d, \, e > 0 \textbf{ in } T_1.$$

Next, we try to determine the range of the two program variables s and q. The assignments to s in program T_1 are $s := 1$ and $s := s/2$; the variable s is initialized to 1 before the loop and is repeatedly divided by 2 within the loop. It follows that $s = 1/2^n$, where n is some nonnegative integer indicating the number of times that s has been halved. Since this relation holds throughout T_1 from the point when the assignment

---

[4]The bug presumably occurred when the programmer "inadvertently" interchanged the two statements within the loop. How a program development system might uncover such bugs is another question; for one possibility, see [KatzManna75a]. [KatzManna76] show how to use invariants to prove the *incorrectness* of a program.

$s := 1$ is first executed, we may assert the additional *global* invariant

assert $(\exists\, n \in \mathbf{N})\, s = 1/2^n$ **in** T_1,

For short, we write

assert $s \in 1/2^{\mathbf{N}}$ **in** T_1,

where $1/2^{\mathbf{N}}$ denotes the set $\{1/2^n : n \in \mathbf{N}\}$. From this invariant one can derive both an upper and lower bound on s. At one extreme, $s = 1/2^0 = 1$, and, at the other extreme (as the exponent increases), the value of s approaches 0. Thus, we may conclude

assert $0 < s \leq 1$ **in** T_1.

The program also contains two assignments to the variable q, $q := 0$ and $q := q + s$. Since we have already determined that s is always of the form $1/2^n$, it follows that q must be a sum of some finite number (possibly zero) of elements of that form. This does not tell too much about q; it does, though, give the lower bound

assert $q \geq 0$ **in** T_1,

since s has been shown to be always positive.

The loop terminates when the exit test $s \leq e$ becomes true. It follows that the relation $s \leq e$ must hold when control reaches the label E_1. This can be asserted in a *local* invariant

E_1: **assert** $s \leq e$.

Similarly, if the exit test is not taken and the loop body is executed, then the exit test must have been false, i.e $s > e$. Neither branch of the conditional statement affects s, and, therefore, the relation $s > e$ holds after the conditional statement as well. At that point s is divided by 2. If before the division we had $s > e$, then at the end of the loop body we have $2 \cdot s > e$. So, whenever the loop body is executed, control returns to the head of the loop with the relation $2 \cdot s > e$ holding. Since that relation does not necessarily hold when the loop is first entered with $s = 1$, it itself is not a loop invariant. Nevertheless, the disjunction of the relations $s = 1$ and $2 \cdot s > e$ is a loop invariant, since one relation holds when the loop is first entered and the other holds every time the loop is repeated, i.e. we have

$$L_1: \textbf{assert } s = 1 \lor 2 \cdot s > e.$$

Turning now to the body of the loop, we consider the conditional statement

$$\textbf{if } d \cdot (q + s) \leq c \textbf{ then } q := q + s \textbf{ fi}.$$

The **then**-path of the conditional statement is taken when $d \cdot (q + s) \leq c$; therefore, after resetting q to $q + s$ we have $d \cdot q \leq c$. Since conditional statements are often intended to achieve the same purpose in different cases, it is plausible that the relation $d \cdot q \leq c$ — achieved by the **then**-path of the conditional—is meant to hold also when the **then**-path is not taken. This suggests the candidate

$$L_1: \textbf{suggest } d \cdot q \leq c.$$

Indeed, since $d \cdot q \leq c$ is true initially, when $q = 0$ and $c \geq 0$, and is unaffected when the conditional test is false (since the value of q is not changed), it invariantly holds when control reaches the head of the loop. We have derived the loop invariant

$$L_1: \textbf{assert } d \cdot q \leq c.$$

The **then**-path is not taken when $c < d \cdot (q + s)$. In that case s is divided in half and q is left unchanged, yielding $c < d \cdot (q + 2 \cdot s)$ at the end of the current iteration. It turns out that this relation holds before the loop is entered and is preserved by the **then**-path. Thus, we have the additional invariant

$$L_1: \textbf{assert } c < d \cdot (q + 2 \cdot s).$$

The loop invariants $d \cdot q \leq c$ and $c < d \cdot (q + 2 \cdot s)$ remain true when the loop exit is taken; along with the exit test $s \leq e$, they imply that, upon termination of the program, the output invariant

$$E_1: \textbf{assert } |c/d - q| < 2 \cdot e$$

holds. Note that the desired relation $|c/d - q| < e$ is *not* implied.

The annotated program—with invariants that correctly express what the program does—is

---

assert $0 \leq c < d$, $e > 0$, $s \in 1/2^N$, $q \geq 0$ **in**
T_1: **begin comment** *annotated buggy real-division program*
  B_1: **assert** $0 \leq c < d$, $e > 0$
  purpose $|c/d - q| < e$
    purpose $q \leq c/d$, $c/d < q + s$, $s \leq e$
      $(q,s) := (0,1)$
      loop L_1: **assert** $d \cdot q \leq c$, $c < d \cdot (q + 2 \cdot s)$, $s = 1 \vee 2 \cdot s > e$
        suggest $c/d < q + s$
        until $s \leq e$
        purpose $q \leq c/d$, $c/d < q + s$, $0 < s < s_{L_1}$
        if $d \cdot (q + s) \leq c$ **then** $q := q + s$ **fi**
        $s := s/2$
      repeat
    assert $q \leq c/d$, $c/d < q + 2 \cdot s$, $s \leq e$
    suggest $c/d < q + s$
  E_1: **assert** $|c/d - q| < 2 \cdot e$
  suggest $|c/d - q| < e$
  end

2.4. Debugging

Now that we know something about what the program does, we can try to debug it. Our task is to find a correction that changes the actual output invariant

$$\textbf{assert } |c/d - q| < 2 \cdot e$$

to the desired output candidate

$$\textbf{suggest } |c/d - q| < e.$$

We go about this by first looking for a way to transform the actual invariant into the desired one; then we try to apply the same transformation to the program, hopefully correcting the error thereby. Accordingly, we would like to transform the insufficient $|c/d - q| < 2 \cdot e$ into the desired $|c/d - q| < e$; we write

$$|c/d - q| < 2 \cdot e \quad \Rightarrow \quad |c/d - q| < e.$$

The obvious difference between the two expressions is that where the first has $2 \cdot e$, the second has just e. So, to transform $|c/d-q| < 2 \cdot e$ into $|c/d-q| < e$, we need only transform

$$2 \cdot e \quad \Rightarrow \quad e,$$

leaving the other symbols unchanged. This may be accomplished by replacing e with $e/2$, i.e. by applying the transformation

$$e \quad \Rightarrow \quad e/2.$$

So far we know that the transformation $e \Rightarrow e/2$, applied to the output invariant $|c/d-q| < 2 \cdot e$, yields the desired output specification $|c/d-q| < e$. That same transformation is now applied to the whole annotated program (excluding the programmer's suggestions). The symbol e appears once in the program text: the exit clause

until $s \leq e$

accordingly becomes

until $s \leq e/2$.

The symbol also appears four times in the invariants; for example, the input assertion $e > 0$ transforms into $e/2 > 0$ which is equivalent to $e > 0$.

The transformed program is

```
assert 0 ≤ c < d, e > 0, s ∈ 1/2ᴺ, q ≥ 0 in
P₁′: begin comment corrected real−division program
    B₁′: assert 0 ≤ c < d, e > 0
    (q,s) := (0,1)
    loop L₁′: assert d·q ≤ c, c < d·(q+2·s), s=1 ∨ 4·s > e
        suggest c/d < q+s
        until s ≤ e/2
        if d·(q+s) ≤ c then q := q+s fi
        s := s/2
        repeat
    E₁′: assert |c/d−q| < e
    end
```

In Appendix 1, it is proved that a transformation such as $e \Rightarrow e/2$ preserves the relation between the program text and invariants, i.e. the transformed assertions are invariants of the transformed program. Appendix 4 contains a trace of the automatic debugging of this program.

In this manner, we have modified the program to achieve the intended result $|c/d-q| < e$. But note that the loop invariant still differs from that suggested by the programmer. The difference between the two is that the programmer intended for $c < d \cdot (q+s)$ to be true, while in fact $c < d \cdot (q+2 \cdot s)$ holds. If desired, this can be remedied by applying the transformation

$$s \quad \Rightarrow \quad s/2.$$

Applying this global transformation affects the five occurrences of s in the program code. The exit clause becomes

until $s/2 \leq e/2,$

or equivalently

until $s \leq e;$

the conditional statement becomes

if $d \cdot (q+s/2) \leq c$ **then** $q := q+s/2$ **fi.**

The assignment statement

$$s := 1$$

transforms into

$$s/2 := 1,$$

which, unfortunately, is not a legal assignment, since an expression appears on the left-hand side. The intent of this illegal statement, however, is to

achieve $s/2=1$ **varying** s

which is the same as

$$\text{achieve } s=2 \text{ varying } s,$$

and may be accomplished by the assignment

$$s := 2.$$

Similarly, the assignment

$$s := s/2$$

gives rise to the goal

$$\text{achieve } s/2=(s'/2)/2 \text{ varying } s,$$

where s' represents the value of the variable s prior to this goal. This is the same as the goal $s=s'/2$ which is achieved by the original assignment

$$s := s/2.$$

At this point, the loop body contains the two statements

$$\text{if } d \cdot (q+s/2) \le c \text{ then } q := q+s/2 \text{ fi}$$
$$s := s/2.$$

Since the expression $s/2$ appears three times in the loop body, this program may be slightly improved by first computing that value, as follows:

$$t := s/2$$
$$\text{if } d \cdot (q+t) \le c \text{ then } q := q+t \text{ fi}$$
$$s := t,$$

where t is a new program variable. Since the conditional statement no longer contains s, the assignment $s := t$ can just as well precede it. But then there is no longer any need for t, as s has the same value. The result is

$$s := s/2$$
$$\text{if } d \cdot (q+s) \le c \text{ then } q := q+s \text{ fi}.$$

Incorporating all of the above changes, our final real division program is

---

assert $0 \le c < d$, $e > 0$, $s/2 \in 1/2^N$, $q \ge 0$ **in**
P_1: **begin comment** *real-division program*
  B_1: **assert** $0 \le c < d$, $e > 0$
  purpose $|c/d - q| < e$
    purpose $q \le c/d$, $c/d < q + s$, $s \le e$
    $(q,s) := (0,2)$
      loop L_1: **assert** $d \cdot q \le c$, $c < d \cdot (q + s)$, $s = 2 \lor 2 \cdot s > e$
      until $s \le e$
      purpose $q \le c/d$, $c/d < q + s$, $0 < s < s_{L_1}$
      $s := s/2$
      if $d \cdot (q + s) \le c$ **then** $q := q + s$ **fi**
      repeat
    assert $q \le c/d$, $c/d < q + s$, $s \le e$
  E_1: **assert** $|c/d - q| < e$
  end

Note that this program is almost the same as the original "buggy" program. It differs in two ways: the two loop-body assignments are interchanged (this presumably was the programmer's error), and s is initialized to 2 rather than 1 (either initialization works).

2.5. Modification

Consider the following specification:

Q_1: **begin comment** *real square-root specification*
  assert $a \ge 0$, $e > 0$
  achieve $|\sqrt{a} - r| < e$ **varying** r
  end

We would like to use the corrected real-division program as a basis for the design of the specified program for computing square-roots. (We assume, of course, that the $\sqrt{}$ operator is not primitive.) To this end, we first compare the specifications of the two programs. The output specification of the division program is

$$\textbf{assert } |c/d - q| < e,$$

while the output specification of the desired program is

$$\textbf{achieve } |\sqrt{a} - r| < e \textbf{ varying } r.$$

The obvious analogy between the two is

$$
\begin{array}{ccc}
q & \Longleftrightarrow & r \\
c/d & \Longleftrightarrow & \sqrt{a},
\end{array}
$$

i.e. where the former specification has q, the other has r, and, where the former has c/d, the latter has $\sqrt{a}$. Thus, to obtain a square-root program from the division program, we need to replace q with r and transform c/d into $\sqrt{a}$. One way to do this (we will see another later) would be via the transformations

$$
\begin{array}{ccc}
q & \Longrightarrow & r \\
d & \Longrightarrow & 1 \\
c & \Longrightarrow & \sqrt{a},
\end{array}
$$

which take c/d into $\sqrt{a}/1 = \sqrt{a}$. Here d is transformed into the identity element of the division operator, leaving c to become $\sqrt{a}$.

Applying this set of transformations to the division program P_1 (annotated with invariants essential for demonstrating its correctness), replaces all occurrences of d with 1 and all occurrences of c with $\sqrt{a}$. The resultant program, after minor simplification, is

---

Q_1': **begin comment** *tentative real square–root program*
  B_1': **assert** $\sqrt{a} < 1$, $e > 0$
  $(r,s) := (0,2)$
  loop L_1': **assert** $r \leq \sqrt{a}$, $\sqrt{a} < r + s$
      until $s \leq e$
      $s := s/2$
      if $r + s \leq \sqrt{a}$ **then** $r := r + s$ **fi**
      repeat
  E_1': **assert** $|\sqrt{a} - r| < e$
  end

This transformed program is guaranteed to satisfy the transformed output specification $|\sqrt{a} - r| < e$; unfortunately, it is unexecutable inasmuch as it contains the nonprimitive function $\sqrt{}$ in the conditional test.

Manipulating programs intelligently naturally requires some knowledge about the subject domain. For example, at this point we need the following fact about square-roots:

fact $u \leq \sqrt{v}$ **is** $u^2 \leq v$ **when** $u \geq 0$, $v \geq 0$,

where u and v are universally quantified, i.e.

$$(\forall u,v) \, (u \geq 0 \wedge v \geq 0 \supset u \leq \sqrt{v} \equiv u^2 \leq v).$$

This fact allows us to replace the problematic test $r + s \leq \sqrt{a}$ with the equivalent $(r + s)^2 \leq a$. It is given that a is nonnegative, as required for the two tests to be equivalent; that $r + s$ is also nonnegative, follows from the loop invariant

$$L_1': \textbf{assert } \sqrt{a} < r + s$$

and the

$$\textbf{fact } 0 \leq \sqrt{u}.$$

There remains an additional problem: a transformed program is only guaranteed to satisfy the output specification for those inputs that satisfy the *transformed* input specification. Unfortunately, the transformed input specification,

$$B_1': \textbf{assert } \sqrt{a} < 1, \, e > 0,$$

is more restrictive than the given input specification $a \geq 0$ for Q_1. We can solve this if we can find an alternative manner, based only on the assumption that $a \geq 0$, by which to initialize the transformed loop invariants $r \leq \sqrt{a}$ and $\sqrt{a} < r + s$ prior to entering the loop. Accordingly, we replace the code preceding the loop with the goal

$$\textbf{assert } a \geq 0, \, e > 0$$
$$\textbf{achieve } r \leq \sqrt{a}, \, \sqrt{a} < r + s \textbf{ varying } r,s.$$

To achieve the first subgoal $r \leq \sqrt{a}$, we can let $r = 0$. Then to achieve the second conjunct $\sqrt{a} < r + s = s$, we can let $s = a + 1$. (This requires additional knowledge about square-roots.) The appropriate assignment is

$$(r,s) := (0, a + 1),$$

completing the square-root program

Q_1: **begin comment** *real square-root program*
   B_1: **assert** $a \geq 0$, $e > 0$
   $(r,s) := (0, a+1)$
   loop L_1: **assert** $r \leq \sqrt{a}$, $\sqrt{a} < r+s$
      until $s \leq e$
      $s := s/2$
      if $(r+s)^2 \leq a$ **then** $r := r+s$ **fi**
      repeat
   E_1: **assert** $|\sqrt{a} - r| < e$
   end

An alternative set of transformations for transforming the specification

$$\textbf{assert } |c/d - q| < e$$

of the division program into that of the square-root program is

$$
\begin{array}{ccc}
q & \Rightarrow & r \\
u/v & \Rightarrow & \sqrt{u} \\
c & \Rightarrow & a,
\end{array}
$$

where by $u/v \Rightarrow \sqrt{u}$ we mean that every occurrence of the (general) division operator $/$ is replaced by the square-root operator applied to what was the numerator. Transformations that involve specific functions such as division, are not, however, guaranteed to yield a correct program, since the program may be based on some property that holds for $/$, but not for $\sqrt{\;}$. Such transformations are heuristic in nature; they only suggest a possible analogy between the two programs. Indeed, when applied to the division program P_1 (**purpose** statements and all), the above transformations yield[5]

---

[5]We do not transform the operator $/2$ into a square-root, since for the purposes of the division program halving is not the same as general division. What needs to be done in the absence of such an assumption is discussed in Chapter 4.

Q_1': **begin comment** *transformed real-division program*
  B_1': **suggest** $0 \leq a < d, e > 0$
  purpose $|\sqrt{a} - r| < e$
    purpose $r \leq \sqrt{a}$, $\sqrt{a} < r + s$, $s \leq e$
      $(r,s) := (0,2)$
      loop L_1': **suggest** $d \cdot r \leq a$, $a < d \cdot (r + s)$
        until $s \leq e$
        purpose $r \leq \sqrt{a}$, $\sqrt{a} < r + s$, $0 < s < s_{L_1'}$
        $s := s/2$
        if $d \cdot (r + s) \leq a$ **then** $r := r + s$ **fi**
      repeat
    suggest $r \leq \sqrt{a}$, $\sqrt{a} < r + s$, $s \leq e$
  E_1': **suggest** $|\sqrt{a} - r| < e$
end

which obviously computes a/d, not $\sqrt{a}$. Since this set of transformations is not correctness-preserving, the asserted invariants have been replaced by suggested candidates.

What must be done in such cases is to review the derivation of the program, expressed by the programmer in **purpose** statements, and see where the analogy breaks down. The purpose of the division program was $|c/d - q| < e$, which transformed into $|\sqrt{a} - r| < e$ as desired. The programmer achieved $|c/d - q| < e$ by splitting it into the conjunction of three subgoals given in the statement

purpose $q \leq c/d$, $c/d < q + s$, $s \leq e$

that appeared in the original program. The last conjunct became the exit test for the loop and the other two became loop invariants. These subgoals transformed into

purpose $r \leq \sqrt{a}$, $\sqrt{a} < r + s$, $s \leq e$,

which indeed imply the transformed goal $|\sqrt{a} - r| < e$. The purpose of the loop body of the division program was

purpose $q \leq c/d$, $c/d < q + s$, $0 < s < s_{L_1}$.

In other words, the loop body reachieves the invariants while making progress towards the exit test by decreasing s.[6] (Recall that s_{L_1} denotes the value of s when last at L_1). The corresponding loop-body subgoal of the transformed program is

$$\textbf{purpose } r \leq \sqrt{a}, \ \sqrt{a} < r + s, \ 0 < s < s_{L_1'}.$$

It is here that the analogy breaks down. The division program first decreases s and then introduces a conditional with the

$$\textbf{purpose } q \leq c/d, \ c/d < q + s$$

for the current value of s. Then it considers two cases, testing if $d \cdot (q+s) \leq c$ or not. If $d \cdot (q+s) \leq c$ does not hold, then $c/d < q+s$, as desired; if $d \cdot (q+s) \leq c$, then $q \leq c/d$ after assigning $q := q+s$, as is also desired. Unfortunately, the fact that the transformed test $d \cdot (r+s) \leq a$ does (or does not) hold, tells nothing about the relation $\sqrt{a} < r+s$ needed for the square-root program. We look, therefore, for a way to transform

$$d \cdot (r+s) > a \ \Rightarrow \ \sqrt{a} < r+s.$$

Since $r+s$ is nonnegative within the program, the right-hand side may be replaced with $(r+s)^2 > a$. Matching the two sides suggests transforming $d \cdot (r+s) \Rightarrow (r+s)^2$. Thus, where the division program has the function $u \cdot v$, the square-root program requires v^2. Accordingly, we complete the analogy by adding the transformation $u \cdot v \Rightarrow v^2$; the full set of transformations now is

$$
\begin{array}{ccc}
q & \Rightarrow & r \\
u/v & \Rightarrow & \sqrt{u} \\
c & \Rightarrow & a \\
u \cdot v & \Rightarrow & v^2 .
\end{array}
$$

As with the first set of transformations, the initialization subgoal does not hold, and must be replaced by the assignment

$$(r,s) := (0, a+1).$$

The same program Q_1 is obtained.

---

[6]To guarantee termination, the decrease cannot be arbitrarily small.

2.6. Abstraction

At this point, we have two programs, P_1 for finding quotients and Q_1 for finding square-roots. Both programs utilize the binary-search technique. It would be nice if one could extract an abstract version of the two programs that captures the essence of the technique, but is not specific to either problem. The resultant abstract program schema could be used as a model of binary search for the solution of future problems.

For that purpose, consider the complete second analogy that we found between the specifications of P_1 and Q_1:

$$
\begin{array}{ccc}
q & \Longleftrightarrow & r \\
u/v & \Longleftrightarrow & \sqrt{u} \\
c & \Longleftrightarrow & a \\
u \cdot v & \Longleftrightarrow & v^2 .
\end{array}
$$

Since both u/v and $\sqrt{u}$ are functions, we try to generalize them to an abstract function $\gamma(u,v)$. Similarly, the generalization of $u \cdot v$ and v^2 should be another function $\delta(u,v)$. Both q and r are output variables and are generalized to an abstract output variable z; the input variables c and a are generalized to an abstract input variable x. This gives the following set of transformations for generalizing the division program:

$$
\begin{array}{ccc}
q & \Longrightarrow & z \\
u/v & \Longrightarrow & \gamma(u,v) \\
c & \Longrightarrow & x \\
u \cdot v & \Longrightarrow & \delta(u,v) .
\end{array}
$$

Applying these transformations to the specification

$$\textbf{achieve } |c/d-q| < e \textbf{ varying } q$$

of the division program yields

$$\textbf{achieve } |\gamma(x,d)-z| < e \textbf{ varying } z.$$

This will be the abstract output specification of the schema. Substituting the abstract functions γ and δ into their respective positions in the annotated division program P_1, we derive the schema

S_1': **begin comment** *abstracted real–division program*
$\quad B_1'$: **suggest** $0 \leq x < d, \, e > 0$
$\quad (z,s) := (0,2)$
$\quad$ **loop** L_1': **suggest** $\delta(d,z) \leq x, \, x < \delta(d,z+s)$
$\qquad\qquad$ **until** $s \leq e$
$\qquad\quad s := s/2$
$\qquad\quad$ **if** $\delta(d,z+s) \leq x$ **then** $z := z+s$ **fi**
$\qquad\quad$ **repeat**
$\quad E_1'$: **suggest** $|\gamma(x,d)-z| < e$
$\quad$ **end**

This schema does not necessarily work for all instantiations of γ and δ, as the original program relied upon facts specific to multiplication and division. Indeed, there is nothing in this abstract program to relate the function γ that appears in the output specification with the function δ that appears in the loop invariant. As we have seen, transformations of specific functions or predicates (such as the suggested abstraction mapping) are *not* necessarily correctness-preserving. We must therefore determine under what conditions this abstract schema does achieve its specifications.

To begin with, the initialization assignment does not achieve the desired loop invariants. We therefore replace the initialization assignment with the subgoal

$\qquad$ **achieve** $\delta(d,z) \leq x, \, x < \delta(d,z+s)$ **varying** z,s,

leaving the specifics of how to initialize the two loop invariants unspecified.

For the loop-body path to be correct, the truth of the suggested invariants must imply that they will hold the next time around; this can easily be shown to be the case for any function δ. For the loop-exit path to be correct, we must have that the loop invariants, plus exit test, imply that the output invariant holds. That is,

$$\delta(d,z) \leq x \wedge x < \delta(d,z+s) \wedge s \leq e \supset |\gamma(x,d)-z| < e.$$

For this to be the case, it suffices to establish the simpler condition[7]

$$\delta(w,u) \leq v \equiv u \leq \gamma(v,w).$$

This condition holds if γ is monotonic [i.e. $v \leq v'$ if and only if $\gamma(v,w) \leq \gamma(v',w)$] and δ is the inverse of γ [i.e. $\gamma(\delta(w,u),w)=u$], as—for example—the function $w \cdot u$ is the inverse of the monotonic function v/w.

In this manner, we have derived the following general program schema for a binary search for the value of $\gamma(x,d)$ within a tolerance e:

S_1: **begin comment** *binary-search schema*
  B_1: **assert** $0 \leq x < d$, $e > 0$, $\delta(w,u) \leq v \equiv u \leq \gamma(v,w)$
  achieve $\delta(d,z) \leq x$, $x < \delta(d,z+s)$ **varying** z,s
  loop L_1: **assert** $\delta(d,z) \leq x$, $x < \delta(d,z+s)$
      until $s \leq e$
      $s := s/2$
      if $\delta(d,z+s) \leq x$ **then** $z := z+s$ **fi**
      repeat
  E_1: **assert** $|\gamma(x,d)-z| < e$
  end

Its abstract output specification is

$$\textbf{assert } |\gamma(x,d)-z| \leq e,$$

and the preconditions for its guaranteed correctness are given in

$$\textbf{assert } 0 \leq x < d, \ e > 0, \ \delta(w,u) \leq v \equiv u \leq \gamma(v,w).$$

Of course, for this schema to be executable, the function δ appearing in it must be primitive; otherwise, it should be replaced. Similarly, the unachieved subgoal

$$\textbf{achieve } \delta(d,z) \leq x, \ x < \delta(d,z+s) \textbf{ varying } z,s$$

must be reduced to primitives.

---

[7]In this condition, u, v, and w are universally quantified. In general, variables that appear in an input specification, but not in the program are taken to be (implicitly) universally quantified.

2.7. Instantiation

The binary-search schema just derived from the division program may be applied to other tasks, such as the computation of the square root of an integer. Our goal now is to use that schema to help construct a program that finds the integer square-root z of a nonnegative integer a:

P_2: **begin comment** *integer square-root specification*
  assert $a \in \mathbf{N}$
  achieve $z = \lfloor \sqrt{a} \rfloor$ **varying** z
 end

where the function $\lfloor u \rfloor$ yields the largest integer not greater than u.

We cannot directly match this goal with the output specification of the schema

$$\textbf{assert } |\gamma(x,d) - z| < e \textbf{ varying } z,$$

or with any of the other invariants known to hold upon termination of the schema. However, if we expand the goal $z = \lfloor \sqrt{a} \rfloor$, using the definition

$$\textbf{fact } v = \lfloor u \rfloor \textbf{ is } v \leq u,\ u < v+1,\ v \in \mathbf{Z}$$

of $\lfloor u \rfloor$ (where $\mathbf{Z}$ is the set of all integers), we get the equivalent goal

$$\textbf{achieve } z \leq \sqrt{a},\ \sqrt{a} < z+1,\ z \in \mathbf{Z} \textbf{ varying } z,$$

i.e. z should be the largest integer not greater than $\sqrt{a}$. Since we know that the schema achieves the two output invariants

$$\textbf{assert } z \leq \gamma(x,d),\ \gamma(x,d) < z + e,$$

we can compare these invariants with the above goal. This suggests the transformations

$$\begin{aligned} \gamma(u,v) &\Rightarrow \sqrt{u} \\ x &\Rightarrow a \\ e &\Rightarrow 1 \end{aligned}$$

to achieve the two conjuncts $z \leq \sqrt{a}$ and $\sqrt{a} < z+1$. In addition, we will have to extend the program to ensure that the final value of z is a nonnegative integer; for this purpose, we append the subgoal

achieve $z \in \mathbf{Z}$ **protecting** $z \leq \sqrt{a}$, $\sqrt{a} < z+1$ **varying** z

to the end of the instantiated schema. The **protecting** clause means that the relations $z \leq \sqrt{a}$ and $\sqrt{a} < z+1$, achieved by the instantiated schema, should not be clobbered when achieving the additional goal $z \in \mathbf{Z}$.

The preconditions for the schema's correctness are

assert $0 \leq x < d$, $e > 0$, $\delta(w,u) \leq v \equiv u \leq \gamma(v,w)$;

instantiating them yields

assert $0 \leq a < d$, $1 > 0$, $\delta(w,u) \leq v \equiv u \leq \sqrt{v}$.

That a is nonnegative is given. To satisfy $a < d$ we could let d be $a+1$. Since

fact $u \leq \sqrt{v}$ **is** $u^2 \leq v$ **when** $u \geq 0$, $v \geq 0$,

the third condition may be satisfied by taking $\delta(w,u)$ to be u^2, provided that the argument u is never negative.[8] This completes the analogy, and suggests the transformations

$$
\begin{aligned}
\gamma(u,v) &\Rightarrow \sqrt{u} \\
x &\Rightarrow a \\
e &\Rightarrow 1 \\
d &\Rightarrow a+1 \\
\delta(w,u) &\Rightarrow u^2.
\end{aligned}
$$

Applying this instantiation to the schema, we obtain the following partially written program:

---

[8]Actually, since the instantiation of $\delta(w,u)$ ignores its first argument, there is no need to instantiate d which only appears in the schema as an argument to δ.

P_2: **begin comment** *instantiated binary–search schema*
$\quad$ B_2: **assert** $a \in \mathbf{N}$
$\quad$ **achieve** $z^2 \leq a$, $a < (z+s)^2$ **varying** z,s
$\quad$ **loop** L_2: **assert** $z^2 \leq a$, $a < (z+s)^2$
$\quad\quad$ **until** $s \leq 1$
$\quad\quad$ $s := s/2$
$\quad\quad$ **if** $(z+s)^2 \leq a$ **then** $z := z+s$ **fi**
$\quad\quad$ **repeat**
$\quad$ **assert** $|\sqrt{a} - z| < 1$
$\quad$ **achieve** $z \in \mathbf{Z}$ **protecting** $z \leq \sqrt{a}$, $\sqrt{a} < z+1$ **varying** z
$\quad$ **end**

That the argument $z + s$ is indeed nonnegative follows from the invariant $\sqrt{a} < z+s$.

2.8. Synthesis

This program still contains two unachieved subgoals:

$\quad\quad$ **achieve** $z^2 \leq a$, $a < (z+s)^2$ **varying** z,s

and

$\quad\quad$ **achieve** $z \in \mathbf{Z}$ **protecting** $z \leq \sqrt{a}$, $\sqrt{a} < z+1$ **varying** z.

We show now how code to achieve these subgoals may be synthesized.

The first subgoal is a conjunction of two relations, $z^2 \leq a$ and $a < (z+s)^2$, which are to be achieved simultaneously. It may be split into the two consecutive subgoals

$\quad\quad$ **purpose** $z^2 \leq a$, $a < (z+s)^2$
$\quad\quad\quad$ **achieve** $z^2 \leq a$ **varying** z
$\quad\quad\quad$ **achieve** $a < (z+s)^2$ **varying** s
$\quad\quad$ **assert** $z^2 \leq a$, $a < (z+s)^2$.

The first sets the variable z to some value satisfying $z^2 \leq a$; the second leaves z constant so that $z^2 \leq a$ remains true, while setting s to some value satisfying $a < (z+s)^2$. Given the

$\quad\quad$ **fact** $u^2 \geq 0$,

we can replace the first subgoal with

achieve $z=0$ **varying** z.

This in turn may be achieved by setting

$$z := 0.$$

Since z is not varied while achieving the second subgoal $a < (z+s)^2$, that subgoal becomes

achieve $a < s^2$ **varying** s.

We shall return to this subgoal later. (Of course, there is no assurance, as yet, that this split will lead eventually to a satisfactory solution; were it not to work out in the end, then we would have to retrace our steps and try something else.)

There are two ways in which we might achieve the other remaining subgoal

achieve $z \in \mathbf{Z}$ **protecting** $z \leq \sqrt{a}$, $\sqrt{a} < z+1$ **varying** z.

One is to take the current value of z which satisfies the two conditions $z \leq \sqrt{a}$ and $\sqrt{a} < z+1$ and perturb it just enough to make it an integer while preserving those two protected relations. This can be done by assigning[9]

if $\lceil z \rceil^2 \leq a$ **then** $z := \lceil z \rceil$ **else** $z := \lfloor z \rfloor$ **fi**.

Alternatively, note that we must actually achieve $z \in \mathbf{N}$, since the protected invariant $\sqrt{a} < z+1$ implies that z is nonnegative. To this end, we set up the goal

achieve $z \in \mathbf{N}$ **in** P_2 **varying** z,

meaning that $z \in \mathbf{N}$ is to be a global invariant of P_2. Accordingly, we must establish $z \in \mathbf{N}$ initially and then preserve it throughout the loop computation. Initially $z=0 \in \mathbf{N}$, as is desired. Since z is sometimes incremented by s, the latter should also be a nonnegative integer. That

---

[9]If $\lceil z \rceil^2 \leq a$ then $\lceil z \rceil \leq \sqrt{a}$ and $\sqrt{a} < z+1 \leq \lceil z \rceil + 1$; on the other hand, if $\lceil z \rceil^2 > a$ then $\lfloor z \rfloor \leq z \leq \sqrt{a}$ and $\sqrt{a} < \lceil z \rceil \leq \lfloor z \rfloor + 1$.

gives a new goal

achieve $s \in N$ in P_2 varying s.

Finally, in order to preserve the invariant $s \in N$, while it is repeatedly halved until it is no longer greater than 1, it is necessary and sufficient that $s \in 2^N$ be invariant, where 2^N denotes the set $\{2^n : n \in N\}$. Thus, we have the stronger goal

achieve $s \in 2^N$ in P_2 varying s.

For $s \in 2^N$ to hold throughout, we also need to ensure that it holds upon entering the loop. Accordingly, we add the conjunct $s \in 2^N$ to the initialization subgoal $a < s^2$.

At this point, we have an unachieved subgoal

achieve $a < s^2$, $s \in 2^N$ varying s.

The first conjunct might be achieved by letting $s = a + 1$, while the second could easily be achieved by letting $s = 1$. However, though each conjunct is achievable by itself in this manner, achieving both together is more difficult, since in general these two solutions conflict. So, instead, we transform this conjunctive goal into an iterative loop,

> **purpose** $a < s^2$, $s \in 2^N$
>   **achieve** $s \in 2^N$ **varying** s
>   **loop** L_2': **assert** $s \in 2^N$
>   **until** $a < s^2$
>   **approach** $a < s^2$ **protecting** $s \in 2^N$ **varying** s
>   **repeat**
> **assert** $a < s^2$, $s \in 2^N$.

choosing first to achieve $s \in 2^N$, and then to keep it true while executing the loop until the remaining conjunct $a < s^2$ is also satisfied.

To initialize $s \in 2^N$, we let $s = 2^0$ and assign

$$s := 1.$$

Within the loop, we have the subgoal

approach $a < s^2$ protecting $s \in 2^N$ varying s,

i.e. we wish to preserve the invariant $s \in 2^N$ while making progress towards the exit test $a < s^2$. Since we know that initially $s = 1$, and, ultimately, we want $0 \leq \sqrt{a} < s$, it follows that s increases within the loop. By presuming that s increases *monotonically*, we get the loop-body subgoal

$$\textbf{achieve } s > s_{L_2'} \textbf{ protecting } s \in 2^{\mathbf{N}} \textbf{ varying } s,$$

where $s_{L_2'}$ denotes the value of s when control was last at the head of this loop. It follows that s must be multiplied by some positive power of two, e.g.

$$s := 2 \cdot s.$$

Putting together all the pieces, we have obtained the following program:

```
assert a ∈ N, z ∈ N, s ∈ 2ᴺ in
P₂: begin comment integer square-root program
    B₂: assert a ∈ N
    z := 0
    purpose a < s², s ∈ 2ᴺ
        s := 1
        loop L₂′: assert s ∈ 2ᴺ
                  until a < s²
                  s := 2·s
                  repeat
    purpose z² ≤ a, a < (z+s)², s ≤ 1
        loop L₂: assert z² ≤ a, a < (z+s)²
                 until s ≤ 1
                 s := s/2
                 if (z+s)² ≤ a then z := z+s fi
                 repeat
    E₂: assert z = ⌊ √a ⌋
    end
```

2.9. Discussion

In this chapter, we have followed one program through many vicissitudes. After correcting a simple but erroneous program, we developed it into a schema and into two other programs. In the chapters that follow, we look at the different aspects of program manipulation in greater depth. In the process, we shall modify the square-root program back again, so that it divides by doubling and halving (in Section 3.3.2).

Chapter 3

Program Modification and Debugging

Although new and important modifications may not arise from reversion and analogous variation, such modifications will add to the beautiful and harmonious diversity of nature.

Modifications in hard parts and in external parts sometimes affect softer and internal parts.

—Charles Darwin (The origin of species)

'Similarity' is not a thing offered on a plate (or hidden in a cupboard); it is a relation established in the mind by a process of selective emphasis on those features which overlap in a certain respect—along one dimensional gradient—and ignoring other features.

—Arthur Koestler (The act of creation)

3.1. Introduction

Distinct species, in general, will have some aspects in common and some distinguishing differences; some parts similar in function but different in form, some different in function but similar in form. Analogously, distinct programs may use similar techniques for disparate tasks, or achieve the same goal in divergent ways. In this chapter, we look into means of modifying one program to accomplish a different goal in a similar manner. An analogy between the specification of the given program and that of the desired one is used as the basis for transforming the existing program to meet the new specification. We also consider debugging in this chapter. Debugging is a special case of modification in which a program is modified to achieve the correct results.

The importance of analogical reasoning has been stressed by many, from Descartes to Pólya. For a discussion of the role of analogy in the sciences, see [Hesse66]; for a review of psychological theories of analogical reasoning, see [Sternberg77]. An early work on automating analogical reasoning is [Evans68]. The use of analogy in automated problem solving in general, and theorem proving in particular, was proposed in [Kling71]. Other works employing analogy as an implement in problem solving include [MooreNewell73], [Brown76], [ChenFindler76], [BrownTarnlund79], [McDermott79], [Winston80], [Carbonell83], and [Burstein83]. The use of analogies to guide the modification of programs was suggested in [MannaWaldinger75] and pursued further in [DershowitzManna77] and [UlrichMoll77].

The idea that programs should be constructed by a series of transformations has been widely promoted (e.g. [Knuth74], [Gerhart75a], [Gerhart75b], [Bauer76], [DarlingtonBurstall76], [StandishKibler-Neighbors76], [Wegbreit76], [Schwartz77], [Loveman77], [Blikle78], [Balzer81], and [BroyPepper81]). Modification differs from such transformations in that correctness with respect to the original specification is *not* preserved. What we want is for the resultant program to be correct with respect to a *transformed* specification. Correctness-preserving transformations and specification-changing modifications are thus complementary.

All our programs are assumed to be annotated with an *output specification* (stating the desired relationship between the input and output variables upon termination of the program), an *input specification* (defining the set of legal inputs on which the program is intended to operate), and *invariant assertions* (relations that are known to always hold at specific points in the program for the current values of variables) demonstrating its correctness. As we shall see, invariant assertions can play an important role during modification.

A scenario of computer-aided programming and debugging appeared in [Floyd71]; one overview of approaches to debugging is [Schwartz71]; some early studies of human debugging activity are [GrantSackman67] and [GouldDrongowski74]. The HACKER system ([Sussman75]) constructed programs by trying out alternatives and attempting to debug them when necessary. [KatzManna75a] and [Sagiv76] describe debugging techniques based—like our method—on invariant assertions. [Boyer-ElspasLevitt75], [Clarke76], [King76], and [Howden77] propose debugging aids based on the symbolic execution of a program. Knowledge-based approaches to debugging include [Stefanesco78], [Wertz79], [AdamLaurent80], [LaubschEisenstadt81], and [Badger,*etal*.82]; plan-based approaches may be found in [MillerGoldstein77] and [Waters82]. Two systems that debug student programs by matching them with the teacher's archetype are [Ruth76] and [Soloway,*etal*.81]

The next section elucidates the basic aspects of our approach to modification with the aid of relatively straightforward examples. More subtle facets of the method are illustrated in the examples of Section 3.3. The correctness of the method is proved in Appendix 1.

3.2. Overview

For program modification, one needs a program satisfying its input-output specification along with the specification for a new program. Comparison of the two specifications may suggest a transformation of the given program that will bring it line with the new specification. Even if the transformed program does not exactly fulfill the goal, it may serve as a basis for constructing the desired program.

3.2.1. Basic Method: Analogy

In this chapter, we stress transformations in which all occurrences of a particular symbol throughout a program are affected. Such transformations are termed "global," in contrast with "local" transformations that are applied only to segments of a program. Local transformations are considered in the next chapter.

As a simple example, consider the following annotated program[1]

P_3: **begin comment** *tournament – minimum program*
   B_3: **assert** $n \geq 0$
   $y := n$
  loop L_3: **assert** $min\{A[y:2 \cdot y]\} = min\{A[n:2 \cdot n]\}$
     until $y = 0$
     if $A[2 \cdot y - 1] \leq A[2 \cdot y]$ **then** $A[y-1] := A[2 \cdot y - 1]$
                 else $A[y-1] := A[2 \cdot y]$ **fi**
     $y := y - 1$
     repeat
   E_3: **assert** $A[0] = min\{A[n:2 \cdot n]\}$
  end

Given an array segment $A[n:2 \cdot n]$ that is nonempty (i.e. n is nonnegative), when this program terminates, $A[0]$ will contain the minimum of the values of the $n+1$ array elements $A[n]$, $A[n+1]$,...,$A[2 \cdot n]$. This output specification is expressed in the final statement[2]

$$E_3: \textbf{assert } A[0] = min\{A[n:2 \cdot n]\}.$$

---

[1]Adapted from [Floyd62].

[2]A more complete specification would also include the requirement that the array elements $A[n:2 \cdot n]$ are unchanged by the program.

That this program satisfies its specification may be proved using the loop invariant

$$L_3: \textbf{assert } min\{A[y:2\cdot y]\}=min\{A[n:2\cdot n]\}.$$

This invariant holds initially when $y=n$, since the two sides of the equality are the same in that case. It is maintained true by the loop body, when y is decremented, since the smaller of the two elements that are excluded from the range $[y:2\cdot y]$ becomes the new $A[y-1]$. The invariant, together with the exit test $y=0$, implies that $A[0] = min\{A[y:2\cdot y]\} = min\{A[n:2\cdot n]\}$ upon termination of the loop, as required by the output specification.

To modify this program to compute the maximum of the array, rather than the minimum, we compare the specification

$$E_3: \textbf{assert } A[0]=min\{A[n:2\cdot n]\}$$

of the given program P_3, with the output specification

$$\textbf{achieve } A[0]=max\{A[n:2\cdot n]\} \textbf{ varying } A$$

of the desired program Q_3. We say that we are looking for an analogy

$$A[0]=min\{A[n:2\cdot n]\} \quad \Longleftrightarrow \quad A[0]=max\{A[n:2\cdot n]\}.$$

The obvious analogy between these two specifications is that one has the function min where the other has max, that is

$$min \quad \Longleftrightarrow \quad max.$$

The analogy $min \Longleftrightarrow max$ suggests that occurrences of the symbol min in the first program be replaced by max. But we cannot simply apply such a transformation

$$min \quad \Rightarrow \quad max.$$

In fact, the symbol min appears nowhere in P_3. Instead, certain properties specific to the function min were used in the construction of the program, properties that do not, however, hold for the new function max. Later, we shall see how such a situation can be dealt with.

In the meantime, we can effect the desired transformation

$$A[0]=min\{A[n:2{\cdot}n]\} \quad \Rightarrow \quad A[0]=max\{A[n:2{\cdot}n]\}$$

in another way. We first look for an equivalent formulation of the specifications that highlights an alternative analogy. For example, the function $max\{A\}$ can be replaced by the equivalent $-min\{-A\}$, where $-A$ denotes the array A with each element negated. Thus, we can transform

$$A[0]=min\{A[n:2{\cdot}n]\} \quad \Rightarrow \quad A[0]=-min\{-A[n:2{\cdot}n]\}$$

instead. Since we are trying to avoid transforming the function min, we would like the right-hand sides of both equalities to begin with the same symbol. Negating both sides of $A[0]=-min\{-A[n:2{\cdot}n]\}$, we get

$$A[0]=min\{A[n:2{\cdot}n]\} \quad \Rightarrow \quad -A[0]=min\{-A[n:2{\cdot}n]\}.$$

Now it is apparant that the transformation

$$A \quad \Rightarrow \quad -A,$$

applied to the output specification of the given program, yields

$$E_3\text{: } \textbf{assert } -A[0]=min\{-A[n:2{\cdot}n]\}$$

which is equivalent to the desired goal $A[0]=max\{A[n:2{\cdot}n]\}$.

Since $A \Rightarrow -A$ is a transformation of the array variable A, and variables have no assumed properties, we can obtain a program guaranteed to satisfy the transformed specification by applying this transformation to the program.[3] The variable A appears within the program text only in the conditional statement

if $A[2{\cdot}y-1] \leq A[2{\cdot}y]$ **then** $A[y-1] := A[2{\cdot}y-1]$
            else $A[y-1] := A[2{\cdot}y]$ **fi**.

Applying the transformation $A \Rightarrow -A$ to this statement yields

---

[3]By applying the same transformation to all occurrences of A in the correctness proof of the original program, a proof for the transformed program is obtained. See Appendix 1.

if $-A[2 \cdot y - 1] \leq -A[2 \cdot y]$ **then** $-A[y-1] := -A[2 \cdot y - 1]$
                         else $-A[y-1] := -A[2 \cdot y]$ **fi**.

The transformed test $-A[2 \cdot y - 1] \leq -A[2 \cdot y]$ is equivalent to $A[2 \cdot y - 1] \geq A[2 \cdot y]$. But the transformed assignment statements are "illegal" since a function, in this case $-A$, may not appear on the left-hand side of an assignment. The intent of the illegal statement,

$$-A[y-1] := -A[2 \cdot y - 1],$$

is for the new value of the expression $-A[y-1]$ to be made equal to the current value of $-A[2 \cdot y - 1]$ by changing the value of the array A. In other words, we wish to achieve the relation given by the goal

achieve $-A[y-1] = -A'[2 \cdot y - 1]$ **varying** A,

where A' denotes the value of the array A before this **achieve** statement. To obtain an assignment to A, we first negate both sides of the equality. The new goal is

achieve $A[y-1] = A'[2 \cdot y - 1]$ **varying** A.

Now we may achieve the desired relation between the new value of A and the old by assigning to A the value of the expression on the other side of the equality:

$$A[y-1] := A[2 \cdot y - 1]$$

(which is what we started out with). Similarly, the transformed assignment

$$-A[y-1] := -A[2 \cdot y]$$

becomes

$$A[y-1] := A[2 \cdot y].$$

In general, if ϵ is some expression containing a variable y, then applying a global transformation

$$y \;\Rightarrow\; \epsilon$$

to an assignment to y

$$y := h(y, z)$$

will result in an illegal statement

$$\epsilon := h(\epsilon,z).$$

If y is the only variable in ϵ, say $\epsilon = f(y)$, where the function f has an inverse f^-, then one way to achieve the desired effect

achieve $f(y) = h(f(y'),z)$ **varying** y

is to apply f^- to both sides of the equality—suggesting the assignment

$$y := f^-(h(f(y),z)).$$

When a global transformation replaces one variable with a function of two or more variables, the situation is more complicated. If ϵ is, say, some function $f(y,z)$, then applying the transformation $y \Rightarrow \epsilon$ to a multiple assignment of the form

$$(y,z) := (g(y,z),h(y,z))$$

gives

$$(\epsilon,z) := (g(\epsilon,z),h(\epsilon,z)).$$

That is, we wish to

achieve $f(y,z) = g(\epsilon',z')$, $z = h(\epsilon',z')$ **varying** y,z,

where $\epsilon' = f(y',z')$ and y' and z' denote the prior values of the variables y and z, respectively. Assuming an inverse f^- of f (in the first argument) is available, i.e. $f^-(f(u,v),v) = u$ for all u and v, then we can achieve the desired relations by assigning

$$(y,z) := (f^-(g(\epsilon,z),h(\epsilon,z)),h(\epsilon,z))$$

instead. Note that the value of $f(y,z)$ changes when *either* y or z is assigned to. Therefore, any assignment $z := h(y,z)$ to z alone is also cause for updating the transformed y, and should be treated as though it was the multiple assignment $(y,z) := (y,h(y,z))$.

Global transformations are applied to the invariants annotating the program as well as to the code. In our case, the loop invariant

$$L_3: \textbf{assert } min\{A[y:2 \cdot y]\} = min\{A[n:2 \cdot n]\}$$

is transformed into

$$L_3: \textbf{assert } min\{-A[y:2 \cdot y]\} = min\{-A[n:2 \cdot n]\},$$

or equivalently

$$L_3: \textbf{assert } max\{A[y:2 \cdot y]\} = max\{A[n:2 \cdot n]\}.$$

At the same time, a transformation that affects an input or output variable will change the output invariant. That is why, for program modification, one looks for a transformation of variables appearing in the output invariants that produces relations that imply the desired output specification.

By applying the transformations suggested by the analogy between the specifications to the given program, the following program for finding the maximum, rather than the minimum, is obtained:

```
Q₃: begin comment tournament-maximum program
  B₃: assert n ≥ 0
  y := n
  loop L₃: assert max{A[y:2·y]} = max{A[n:2·n]}
        until y = 0
        if A[2·y - 1] ≥ A[2·y]  then  A[y-1] := A[2·y - 1]
                                else  A[y-1] := A[2·y]      fi
        y := y - 1
        repeat
  E₃: assert A[0] = max{A[n:2·n]}
  end
```

Note that the array $-A$ no longer appears in the program; only the original A is actually used.

3.2.2. Special Case: Debugging

An incorrect program that computes wrong results must be modified to compute the desired, correct results. If one knows what the "bad" program really does, then it may be compared with the specification of what it should do. The result of the comparison may then be used to modify, that is, *debug,* the erroneous program.

As an example, consider the incorrect program[4]

T_4: **begin comment** *given integer square–root program*
$(z,u,v) := (1,0,3)$
loop until $a < u$
    $(z,u,v) := (z+1, u+v, v+2)$
    repeat
E_4: **suggest** $z^2 \leq a$, $a < (z+1)^2$, $z \in \mathbf{N}$
end

---

intended to compute the integer square-root z of the nonnegative integer a. The goal, then, was to set z be just less than the square-root of a, so that a lies between the squares of the integers z and $z+1$:

achieve $z^2 \leq a$, $a < (z+1)^2$, $z \in \mathbf{N}$ **varying** z,

where $\mathbf{N}$ is the set of nonnegative integers. The cause of the bug was the "accidental" exchange of the initial values of z and u.

Using the annotation methods described in Chapter 6, invariants can be generated to express how the above program works. In particular, the global invariants

assert $v = 2 \cdot z + 1$, $u = z^2 - 1$, $z \in \mathbf{N} + 1$ **in** T_4

where $\mathbf{N} + 1$ denotes the set of positive integers, can be shown to hold throughout the program. Also, the invariant $a \geq u - v$ holds each time control reaches the head of the loop, since the exit test $a < u$ must have been false for the loop not to have been exited the previous time around (before u was incremented by v). Combining these invariants with the fact that $a < u$ when the loop terminates, demonstrates that the given program halts with the relations

E_4: **assert** $(z-1)^2 \leq a+1$, $a + 1 < z^2$, $z \in \mathbf{N} + 1$

holding between the output variable z and input variable a. This is to be contrasted with the desired relations

achieve $z^2 \leq a$, $a < (z+1)^2$, $z \in \mathbf{N}$.

Comparing the two sets of relations, we note that the desired goal $z^2 \leq a$

---

[4]Adapted from [MannaNessVuillemin72].

cab be obtained from the actual output relation $(z-1)^2 \leq a+1$ by replacing z with $z+1$ and a with $a-1$:

$$z \;\Rightarrow\; z+1$$
$$a \;\Rightarrow\; a-1.$$

By the same set of transformations, $a < (z+1)^2$ may be obtained from $a+1 < z^2$.

In general: in order to transform an expression of the form $f(s)$ into another expression ϵ, where f is some function, we try to transform $s \Rightarrow f^-(\epsilon)$, where f^- is the inverse function of f, i.e. $f(f^-(x))=x$ for all x, if such an inverse exists. Thus, to transform $z-1$ into z, we replace z with $z+1$; to transform $a+1$ into a, we replace a with $a-1$.

Applying the transformation $a \Rightarrow a-1$ to the program statements affects only the exit test

until $a < u$,

which becomes $a-1 < u$, or equivalently

until $a \leq u$

(a and u being integers). The transformation $z \Rightarrow z+1$ affects two other statements: the initialization

$$z := 1$$

becomes

$$z+1 := 1$$

and the loop-body assignment

$$z := z+1$$

becomes

$$z+1 := z+2.$$

These resultant assignments, however, are "illegal," inasmuch as an expression such as $z+1$ may not appear on the left-hand side of an assignment. Instead, the expression $z+1$ is given the initial value 1 by assigning

$$z := 0,$$

and the value of the expression $z+1$ is incremented to $z+2$ by the "legal" assignment

$$z := z+1.$$

These changes give us the following correct program:

Q_4: **begin comment** *integer square-root program*
$(z,u,v) := (0,0,3)$
loop until $a \leq u$
$\qquad (z,u,v) := (z+1, u+v, v+2)$
$\qquad$ **repeat**
E_4: **assert** $z^2 \leq a$, $a < (z+1)^2$, $z \in \mathbf{N}$
end

In [KatzManna76] it is suggested that when there is insufficient information to prove either the correctness or incorrectness of a program, it may nevertheless be desirable to "debug" it. The possibly incorrect program may be transformed, using known invariants, into a program that is unquestionably correct. Even when invariants are found for an incorrect program—it is often more difficult to discover invariants for an incorrect program since it may in fact not compute anything meaningful—there may be no way to transform those invariants into the desired specification. It then becomes necessary to apply localized transformations by considering where in the program the offending invariants come from.[5]

3.2.3. Correctness Considerations

In the above examples, the transformed programs were correct in the sense that the transformed programs do in fact satisfy the transformed specification. As noted, this is not necessarily the case with

---

[5]This is like the use of "derivation trees" for debugging, suggested in [KatzManna76] and [MillerGoldstein77]. Localized methods are discussed in the next chapter. Some suggestions as to how one might automate the process of proving that a given program is incorrect may be found in [KatzManna75a].

any transformation. Suppose, for example, that one is given the program

---

P_5: **begin comment** *minimum–value program*
  $(y,z) := (0,A[0])$
  loop L_5: **assert** $z=min\{A[0{:}y]\}$
    until $y=n$
    $y := y+1$
    if $A[y] < z$ **then** $z := A[y]$ **fi**
    repeat
  E_5: **assert** $z=min\{A[0{:}n]\}$
  end

---

for finding the minimum of the array $A[0{:}n]$, and wishes to construct a program to find the maximum of the nonempty array $A[1{:}n]$. The given program achieves the output relation

$$E_5: \textbf{assert } z=min\{A[0{:}n]\},$$

while the output specification of the desired program is

$$\textbf{achieve } z=max\{A[1{:}n]\} \textbf{ varying } z.$$

Thus, the transformations $min \Rightarrow max$ and $0 \Rightarrow 1$ suggest themselves. Applying these transformations yields

---

Q_5': **begin comment** *suggested maximum–value program*
  $(y,z) := (1,A[1])$
  loop L_5': **suggest** $z=max\{A[1{:}y]\}$
    until $y=n$
    $y := y+1$
    if $A[y] < z$ **then** $z := A[y]$ **fi**
    repeat
  E_5': **suggest** $z=max\{A[1{:}n]\}$
  end

To signify that correctness is not guaranteed in this case, we have replaced the assertions with suggestions that merely state what the analogous invariants are. Indeed the transformed program is incorrect, since the loop body does not preserve the candidate loop invariant

$$L_5': \textbf{suggest } z=max\{A[1{:}y]\}.$$

We should not have expected the proposed transformation $min \Rightarrow max$ to have a real effect on the program, as min appears nowhere in the code.

To delineate when a transformation is sure to work, we need to distinguish between two types of symbols: a *constant* is any symbol appearing in a program with assumed properties, e.g. 0, **true**, $+$, and $\geq$; a *variable* is any symbol appearing in the program with no assumed properties other than those mentioned in the input specification. Variables that change value during program execution are termed *program variables;* any other variables are considered to be *input variables*. Those program variables that appear in the output specification are *output variables*. For example, the symbols A and n appearing in program P_5 are input variables, z is an output variable, and y is a program variable; the symbols 0, min, $+$, etc. are constants. Global transformations such as $A \Rightarrow -A$, whereby an input variable is replaced by a function of only input variables, or an output variable by a function of output (and perhaps input) variables, will yield a program satisfying the transformed specification. Like simple "changes of variables," such transformations preserve the "meaning" of the program, when applied systematically. However, transformations of constant symbols, such as $min \Rightarrow max$, are not guaranteed to result in a program satisfying the specification. (More details regarding correctness-preserving transformations are provided in Appendix 1.)

Hence, for some transformations, correctness must be verified. For this purpose, invariant assertions are utilized. If the new program is not correct, then the more information about the logic of the program that is available, the easier it will be to find additional changes that might lead to a correct program. Invariants are essential in our approach to debugging too, as it is necessary to have some idea of what the program actually does before it can be corrected.

3.2.4. Completing an Analogy

As discussed above, a transformed program is not necessarily correct. There could, for example, be unrelated occurrences of constants (in which case a global transformation would be inappropriate), or a constant appearing in the specification might not appear explicitly in the program at all (in which case its transformation would be ineffectual). It is therefore necessary to check the validity of transformations of constants.

A program annotated with invariants is correct with respect to its specification if each of its *verification conditions* is valid. Each condition corresponds to a path in the program. For example, the transformed program Q_5' has two loop-body paths, one for the case when the conditional test is true, and one for the false case. The path for the true case may be represented as

$$\textbf{assert } z = max\{A[1:y]\}$$
$$\textbf{assert } y \neq n$$
$$y := y + 1$$
$$\textbf{assert } A[y] < z$$
$$z := A[y]$$
$$\textbf{suggest } z = max\{A[1:y]\}.$$

We assume that the loop invariant $z = max\{A[1:y]\}$ holds before the loop body is executed. We need to show that if that path is taken, that is, if the exit test $y = n$ is false, and if after incrementing y the conditional test $A[y] < z$ is true, then the invariant again holds, for the incremented y and for the new value $A[y]$ of z. To obtain the corresponding verification condition, the assertions and suggestions are "pushed back" over the assignments.[6] For $z = max\{A[1:y]\}$ to be true after assigning $A[y]$ to z, it must be that $A[y] = max\{A[1:y]\}$ before. Asserting $A[y] < z$ after incrementing y is the same as asserting $A[y+1] < z$ before. Similarly, suggesting that $A[y] = max\{A[1:y]\}$ after incrementing y is like suggesting $A[y+1] = max\{A[1:y+1]\}$ before. Thus the verification condition for this path is

---

[6]Appendix 4 includes "backward" rules for that purpose. We do the same informally here.

> **assert** $z=max\{A[1:y]\}$
> **assert** $y\neq n$
> **assert** $A[y+1]<z$
> **suggest** $A[y+1]=max\{A[1:y+1]\}$,

where the assertions should imply the suggestions, i.e.

$$z=max\{A[1:y]\} \wedge y\neq n \wedge A[y+1]<z \supset A[y+1]=max\{A[1:y+1]\}.$$

Unfortunately, this condition does not hold; we must try to find a way to correct it.

The suggestion

> **suggest** $A[y+1]=max\{A[1:y+1]\}$

is equivalent to

> **suggest** $A[y+1]=max(max\{A[1:y]\},A[y+1])$,

which, in turn, simplifies to

> **suggest** $A[y+1]=max(z,A[y+1])$,

given that $z=max\{A[1:y]\}$. The latter suggestion is equivalent to

> **suggest** $A[y+1]\geq z$.

Now, comparing this suggestion with the assertion

> **assert** $A[y+1]<z$

raises the possibility that the analogy can be completed with an additional transformation

$$< \implies \geq.$$

The transformation $< \implies \geq$ in fact validates the verification conditions and yields the following correct program for finding the maximum:

Q_5: **begin comment** *maximum-value program*
   $(y,z) := (1,A[1])$
   loop L_5: **suggest** $z=max\{A[1{:}y]\}$
      until $y=n$
      $y := y+1$
      if $A[y] \geq z$ **then** $z := A[y]$ **fi**
      repeat
   E_5: **assert** $z=max\{A[1{:}n]\}$
   end

Note that the additional transformation need only have been applied to the conditional statement, since its verification conditions are the only ones that fail.

Were 0 not to appear explicitly in the initialization of P_5, say if we had instead

$$(y,z) := (1-1,A[1-1]),$$

then applying the transformations

$$\begin{array}{ccc} 0 & \Rightarrow & 1 \\ min & \Rightarrow & max \end{array}$$

would have no effect and the verification condition

suggest $A[1-1]=max\{A[1{:}1-1]\}$

for the initialization path would not hold. In a situation such as this, it may be necessary to write a new program segment to initialize the loop invariant $z=max\{A[1{:}y]\}$ by setting y and z to appropriate values. That goal,

achieve $z=max\{A[1{:}y]\}$ **varying** $y,z,$

is satisfied, for example, by the assignment

$$(y,z) := (1,A[1]).$$

Were 0 to appear in unrelated parts of the program—say, for the purpose of illustration, that we had an additional loop-body assignment $y := y+0$—then the transformation $0 \Rightarrow 1$ would result in an incorrect program. In such cases, analysis of the (loop-body) verification conditions would suggest not applying that transformation to that occurrence of 0.

Another problem arises in program modification when transformations only achieve part of the output specification. It, nevertheless, may be possible to extend the program to achieve the complete specification by achieving the missing parts at the onset and then maintaining them invariantly true until program termination. Alternatively, one could append new code to the end of the program that achieves the additional parts—without "clobbering" what has already been achieved by the program.

For example, consider the case where it is desired that Q_5 also find the position x, in the array, of the minimum element z. We can extend the program to

achieve $z = A[x]$ in Q_5 varying x

by maintaining the relation $z = A[x]$ as an invariant throughout the execution of the program. Less efficiently, but equally correct, a loop to search for z in A can be synthesized and appended to the end of the program. Synthesis and extension are treated in Chapter 5.

3.3. Examples

In this section, we present three examples of program modification. We begin with an incorrect real-division program and show how it can be corrected. This is the same program as appeared in the overview chapter; here we go into greater detail. (See Appendix 5 for a trace of its automatic debugging.) The second example is the modification of a square-root program for integers to perform integer division. Lastly, we modify a square-root program for reals to search a sorted array for a given element.

3.3.1. Real Division

Consider the problem of computing the quotient q of two nonnegative real numbers c and d within a specified positive tolerance e, where it is given that $c < d$, and suppose that someone wrote the following program to solve that problem:

```
T₁: begin comment given real-division program
  B₁: assert 0 ≤ c < d, e > 0
  (q,s) := (0,1)
  loop  L₁: suggest q ≤ c/d, c/d < q + s
        until s ≤ e
        if d·(q + s) ≤ c then q := q + s fi
        s := s/2
        repeat
  E₁: suggest q ≤ c/d, c/d < q + e
  end
```

The initial assertion

$$B_1: \textbf{assert } 0 \leq c < d, \, e > 0$$

contains the input specification that the input variables c, d, and e are assumed to satisfy. The statement

$$E_1: \textbf{suggest } q \leq c/d, \, c/d < q + e$$

at the end of the program expresses the output specification that the program is meant to achieve. The program, however, is *incorrect*. For example, the input values $c=1$, $d=3$, and $e=1/3$, satisfy the input specification, but the output value $q=0$ does not satisfy the second conjunct $c/d < q + e$ of the output specification.

Before we can debug this program, we must know more about what it actually does. The annotated program—with loop and output invariants that correctly express what the program does[7] —is

---

[7]See Section 2.2.

B_1: **assert** $0 \leq c < d, \, e > 0$
$(q,s) := (0,1)$
loop L_1: **assert** $q \leq c/d, \, c/d < q + 2 \cdot s$
    until $s \leq e$
    if $d \cdot (q + s) \leq c$ **then** $q := q + s$ **fi**
    $s := s/2$
    repeat
E_1: **assert** $q \leq c/d, \, c/d < q + 2 \cdot e$

Note that the desired relation $c/d < q + e$ is *not* implied by the output invariants.

3.3.1.1. First Correction

We now have the task of finding a transformation (correction) that transforms the actual output invariant

$$\textbf{assert } q \leq c/d, \, c/d < q + 2 \cdot e$$

into the desired goal

$$\textbf{suggest } q \leq c/d, \, c/d < q + e.$$

The same transformation will then be applied to the program. Accordingly, we would like to modify the program in such a manner as to transform the insufficiently strong $c/d < q + 2 \cdot e$ into the desired $c/d < q + e$ and at the same time preserve the correctness of the other conjunct of the specification:

$$c/d < q + 2 \cdot e \quad \Rightarrow \quad c/d < q + e$$
$$q \leq c/d \quad \Rightarrow \quad q \leq c/d.$$

The expressions $c/d < q + 2 \cdot e$ and $c/d < q + e$ differ in that the former has $2 \cdot e$ where the latter has just e. So, if we can transform

$$2 \cdot e \quad \Rightarrow \quad e,$$

then we will have transformed the specification as desired. Since halving is the inverse of doubling, in order to transform the expression $2 \cdot e$ into e, we can transform the input variable e into $e/2$. We therefore apply the transformation

$$e \;\Rightarrow\; e/2$$

to all occurrences of e in the program; all other symbols in the program are left unchanged. There is only one executable statement—the exit clause—affected by this transformation. The suggested correction, then, is

> ### *Correction 1*
> Replace the exit clause with
>
> **until** $s \leq e/2$

The transformation $e \Rightarrow e/2$ must be applied to the input assertion as well as to the rest of the annotated program, and could change the range of legal inputs thereby. In our case, the transformation is innocuous, since the transformed input assertion,

$$B_1: \textbf{assert } 0 \leq c < d,\ e/2 > 0,$$

is equivalent to the given input specification

$$B_1: \textbf{assert } 0 \leq c < d,\ e > 0.$$

Thus, halving e has no effect on the input range, and the transformed program

```
P₁': begin comment corrected real-division program
  B₁': assert 0 ≤ c < d, e > 0
  (q,s) := (0,1)
  loop  L₁': assert q ≤ c/d, c/d < q + 2·s
        until s ≤ e/2
        if d·(q+s) ≤ c then q := q+s fi
        s := s/2
        repeat
  E₁': assert q ≤ c/d, c/d < q+e
  end
```

is correct for any inputs satisfying the given specification.

3.3.1.2. Second Correction

Two additional debugging transformations are possible for the given program. For the first, we try to leave e unchanged. Accordingly, we isolate e on the right-hand side of the inequality and try to transform

$$(c/d-q)/2 < e \quad \Rightarrow \quad c/d-q < e.$$

Now the right-hand sides match, leaving $(c/d-q)/2 \Rightarrow c/d-q$. Doubling, we get $c/d-q \Rightarrow 2\cdot(c/d-q)$, or equivalently, $c/d-q \Rightarrow 2\cdot c/d-2\cdot q$. This leads either to the transformations

$$
\begin{aligned}
c &\Rightarrow 2\cdot c \\
q &\Rightarrow 2\cdot q,
\end{aligned}
$$

or to

$$
\begin{aligned}
d &\Rightarrow d/2 \\
q &\Rightarrow 2\cdot q.
\end{aligned}
$$

Applying the first set of transformations to the output assertion $q \le c/d$ gives $2\cdot q \le 2\cdot c/d$, while the second set of transformations gives $2\cdot q \le c/(d/2)$. In both cases the transformed assertion simplifies back to $q \le c/d$, which is exactly what is wanted. No further transformations are necessary.

Each of these two possible sets of transformations involves one of the input variables, and should, therefore, also be applied to the input assertion. Fortunately, these transformations pose no problem, but only because the weaker condition $c < 2\cdot d$ suffices to imply that the loop invariants $d\cdot q \le c$ and $c < d\cdot(q+2\cdot s)$ hold after the initialization assignment $(q,s) := (0,1)$. (This is easily seen by substituting 0 and 1 for q and s, respectively, in the invariants.) Thus we can relax the input assertion of T_1 to

$$B_1: \textbf{assert } 0 \le c < 2\cdot d, \ e > 0,$$

and not insist that $c < d$. Then, replacing the c in $c < 2\cdot d$ by $2\cdot c$ (or the d by $d/2$) yields a program correct for inputs satisfying $c < d$, as is required.

Doubling q and either doubling c or halving d in the conditional test $d\cdot(q+s) \le c$ yields a test equivalent to $d\cdot(q+s/2) \le c$.

Transforming q into $2 \cdot q$ affects two additional statements: the initialization $q := 0$ becomes the "illegal" assignment $2 \cdot q := 0$, which is equivalent to the "legal"

$$q := 0.$$

Similarly, the assignment $q := q + s$ of the **then**-branch becomes $2 \cdot q := 2 \cdot q + s$; in order to achieve the relation $2 \cdot q = 2 \cdot q' + s$ between the old and new values of q, changing only q, we can assign

$$q := q + s/2.$$

No other statements are affected by either of the two modifications; thus, they both yield

Correction 2
Replace the conditional statement with
if $d \cdot (q + s/2) \leq c$ **then** $q := q + s/2$ **fi**

Our program after Correction 2, annotated with appropriately modified invariants, is (all c have been replaced by $2 \cdot c$ and all q by $2 \cdot q$, and the resultant expressions have been simplified)

```
P₁: begin comment real-division program
  B₁: assert 0 ≤ c < d, e > 0
    (q,s) := (0,1)
    loop L₁: assert q ≤ c/d, c/d < q + s
      until s ≤ e
      if d·(q + s/2) ≤ c then q := q + s/2 fi
      s := s/2
    repeat
  E₁: assert q ≤ c/d, c/d < q + e
  end
```

The loop body can be simplified by eliminating the occurrences of the expression $s/2$, as described in Section 2.4.

3.3.2. Integer Division

Suppose we wish to construct a program to compute the quotient q and remainder r of two integers c and d. The formal specification is

---

Q_2: **begin comment** *integer–division specification*
  assert $c \in \mathbb{N}$, $d \in \mathbb{N}+1$
  achieve $q \leq c/d$, $c/d < q+1$, $q \in \mathbb{N}$, $r = c - d \cdot q$ **varying** q, r
  end

Of course, we mean for the general division operator not to be a primitive; better yet, we would prefer that the program not use multiplication by other than a constant.

3.3.2.1. Modification

To achieve the above goal, we will try to modify the following program, developed in Section 2.8:

---

assert $a \in \mathbb{N}$, $z \in \mathbb{N}$, $s \in 2^{\mathbb{N}}$ **in**
P_2: **begin comment** *integer square–root program*
  B_2: **assert** $a \in \mathbb{N}$
  $(z,s) := (0,1)$
  loop L_2': **assert** $s \in 2^{\mathbb{N}}$
        until $a < s^2$
        $s := 2 \cdot s$
        repeat
  loop L_2: **assert** $z^2 \leq a$, $a < (z+s)^2$
        until $s \leq 1$
        $s := s/2$
        if $(z+s)^2 \leq a$ **then** $z := z+s$ **fi**
        repeat
  E_2: **assert** $z = \lfloor \sqrt{a} \rfloor$
  end

This program approximates the square-root of a by finding the largest integer z that is not greater than the exact value of $\sqrt{a}$. The first step is to attempt to discover an analogy between the specifications of the two programs. In this case the output specification

$$E_2: \text{ assert } z = \lfloor \sqrt{a} \rfloor$$

of the square-root program bears little external resemblance with our goal

achieve $q \leq c/d$, $c/d < q+1$, $q \in \mathbf{N}$, $r = c - d \cdot q$ **varying** q,r.

But, if we also know that the output invariants

$$E_2: \text{ assert } z \leq \sqrt{a}, \ \sqrt{a} < z+1, \ z \in \mathbf{Z}$$

($\mathbf{Z}$ is the set of all integers) hold, then the analogy is more readily apparant. One possible set of transformations is

$$
\begin{array}{ccc}
z & \Rightarrow & q \\
\sqrt{a} & \Rightarrow & c/d.
\end{array}
$$

The latter transformation can be accomplished by letting

$$a \ \Rightarrow \ (c/d)^2.$$

These transformations will achieve the desired conjuncts $q \leq c/d$ and $c/d < q+1$; in addition we will have to extend the program to achieve $r = c - d \cdot q$.

Applying these transformations, the exit test of the first loop, $a < s^2$, becomes $(c/d)^2 < s^2$. Since c, d, and s are all positive, this is the same as $c/d < s$ or $c < d \cdot s$. Similarly, the conditional test $(z+s)^2 \leq a$ becomes $d \cdot (q+s) \leq c$. The transformed program accordingly is

```
(q,s) := (0,1)
loop L₂': assert s ∈ 2ᴺ
          until c < d·s
          s := 2·s
          repeat
loop L₂: assert q ≤ c/d, c/d < q+s
          until s ≤ 1
          s := s/2
          if d·(q+s) ≤ c then q := q+s fi
          repeat
```

By applying the transformation $a \Rightarrow (c/d)^2$ to the given input specification $a \in \mathbf{N}$ of the integer square-root program, the condition

$(c/d)^2 \in N$ on the inputs c and d is obtained. The problem is that this is stronger than the specification $c \in N$ and $d \in N+1$ for integer division. However, all that is actually needed for the correctness of the square-root program is $a \geq 0$; the input specification $a \in N$ is unnecessarily restrictive. Applying the transformation to $a \geq 0$ instead, yields $(c/d)^2 \geq 0$. Since this is implied by $c \in N$ and $d \in N+1$, the abov program is correct for all legal inputs.

3.3.2.2. Extension

At this point, we have a program that achieves three out of the four requirements in the output specification of Q_2. To achieve the additional requirement $r = c - d \cdot q$, the above program can be extended by incorporating the output variable r and maintaining the desired relation true throughout the program. This requires that whenever q is assigned to, r be updated in tandem. Thus, when q is initialized to 0, r is set to $c - d \cdot 0 = c$. When q is incremented by s, r is updated to $c - d \cdot (q + s) = r - d \cdot s$. (Extension is discussed in greater detail in Chapter 5.)

With the new assignments to r included, we get the program

```
(q,s,r) := (0,1,c)
loop L₂': assert s ∈ 2ᴺ
          until c < d·s
          s := 2·s
          repeat
loop L₂: assert q ≤ c/d, c/d < q+s
         until s ≤ 1
         s := s/2
         if d·(q+s) ≤ c then (q,r) := (q+s,r-d·s) fi
         repeat.
```

Note that the conditional test $d \cdot (q + s) \leq c$, which is equivalent to $d \cdot s \leq c - d \cdot q$, can be replaced by $d \cdot s \leq r$, now that r is available. Furthermore, the expression $d \cdot s$ involves multiplication and appears several times. By introducing a new variable u and extending the program once more to maintain the new relation $u = d \cdot s$ globally invariant, we can improve the program. Substituting u for all occurrences of $d \cdot s$ and updating it whenever the value of s is changed, we obtain

---

assert $c \in \mathbb{N}$, $d \in \mathbb{N}+1$, $q \in \mathbb{N}$, $s \in 2^{\mathbb{N}}$, $r = c - d \cdot q$, $u = d \cdot s$ **in**
Q_2: **begin comment** *integer-division program*
   B_2: **assert** $c \in \mathbb{N}$, $d \in \mathbb{N}+1$
   $(q,s,r,u) := (0,1,c,d)$
   loop L_2': **assert** $s \in 2^{\mathbb{N}}$
      until $c < u$
      $(s,u) := (2 \cdot s, 2 \cdot u)$
      repeat
   loop L_2: **assert** $q \leq c/d$, $c/d < q+s$
      until $s \leq 1$
      $(s,u) := (s/2, u/2)$
      if $u \leq r$ **then** $(q,r) := (q+s, r-u)$ **fi**
      repeat
   E_2: **assert** $q \leq c/d$, $c/d < q+1$, $q \in \mathbb{N}$, $r = c - d \cdot q$
   end

This is the desired "hardware" integer-division program. Its only operations are addition, subtraction, comparison, and shifting—hardware instructions on binary computers.

3.3.3. Array Search

In this example, we show how the square-root program

Q_1: **begin comment** *square-root program*
   B_1: **assert** $a \geq 1$, $e > 0$
   $(r,s) := (1, a-1)$
   loop L_1: **assert** $r \leq \sqrt{a}$, $\sqrt{a} < r+s$
      until $s \leq e$
      $s := s/2$
      if $(r+s)^2 \leq a$ **then** $r := r+s$ **fi**
      repeat
   E_1: **assert** $r \leq \sqrt{a}$, $\sqrt{a} < r+e$
   end

---

may be modified to obtain a program that searches for the position z of an element b known to occur in an array segment $A[1{:}n]$. The array is assumed to contain integers sorted in nondescending order. This example will illustrate some of the difficulties that may be encountered in the process of modifying programs.

3.3.3.1. Modification

Our goal is

R_1: **begin comment** *array-search specification*
    assert $u \leq v \supset A[u] \leq A[v]$, $A[u] \in \mathbf{Z}$, $b \in \{A[1{:}n]\}$
    achieve $A[z]{=}b$ **varying** z
    end

where $b \in \{A[1{:}n]\}$ means that the element b occurs in (the multiset of elements of) the array segment $A[1{:}n]$. We are given that A is an array of nondecreasing integers; this is expressed in the input conditions $A[u] \in \mathbf{Z}$ and $u \leq v \supset A[u] \leq A[v]$, where the variables u and v are universally quantified, i.e. each element $A[u]$ of A is in the set of integers $\mathbf{Z}$, and, if the position u is less than or equal to position v, then the value $A[u]$ is less than or equal to the value $A[v]$. Arrays may be indexed by any real number, and we adopt the convention that the intended element may be found by truncating the index,[8] i.e.

$$\textbf{fact } A[u]{=}A[\lfloor u \rfloor]$$

for all u. The desired goal

$$\textbf{achieve } A[z]{=}b \textbf{ varying } z$$

is not directly comparable with the output invariants

$$\textbf{assert } r \leq \sqrt{a}, \; \sqrt{a} < r + e$$

of the given program. So we must first decompose the goal somewhat.

As a first try, we replace the desired goal with the equivalent conjunctive goal

---

[8]In a similar manner, one could develop a program following the Algol-60 convention of rounding-off the index.

achieve $A[z] \leq b$, $b \leq A[z]$ **varying** z,

guided by the fact that we wish to achieve an equality, while the given program achieves an inequality. Since we are dealing with integers, this is the same as

achieve $A[z] \leq b$, $b < A[z]+1$ **varying** z.

Accordingly, we look for a transformation

$$r \leq \sqrt{a} \wedge \sqrt{a} < r+e \ \Rightarrow \ A[z] \leq b \wedge b < A[z]+1.$$

and try to compare the conjunct $r \leq \sqrt{a}$ with $A[z] \leq b$.[9] Matching the two sides of the inequality, we get

$$\begin{array}{ccc} r & \Rightarrow & A[z] \\ \sqrt{a} & \Rightarrow & b. \end{array}$$

To obtain $\sqrt{a} \Rightarrow b$, we can let

$$a \ \Rightarrow \ b^2.$$

Applying these transformations to the remaining conjunct $\sqrt{a} < r+e$ leaves $b < A[z]+e \ \Rightarrow \ b < A[z]+1$, suggesting the additional transformation

$$e \ \Rightarrow \ 1.$$

Applying the three transformations

$$\begin{array}{ccc} r & \Rightarrow & A[z] \\ a & \Rightarrow & b^2 \\ e & \Rightarrow & 1 \end{array}$$

to the given square-root program yields

---

[9]Since $\wedge$ is commutative, one could just as well begin by trying to compare $r \leq \sqrt{a}$ with $b < A[z]+1$.

R_1: **begin comment** *proposed array-search program*
   B_1: **assert** $b^2 \geq 1$, $1 > 0$
   $(A[z], s) := (1, b^2 - 1)$
   loop L_1: **assert** $A[z] \leq b$, $b < A[z] + s$
          until $s \leq 1$
          $s := s/2$
          if $(A[z] + s)^2 \leq b^2$ **then** $A[z] := A[z] + s$ **fi**
          repeat
   E_1: **assert** $A[z] \leq b$, $b < A[z] + 1$
   end

There are, however, a number of problems with this program, the insurmountable one lying in the **then**-branch assignment $A[z] := A[z] + s$. The problem is that the original goal stated that only z is an output variable, while the array A is an input variable which may not be modified by an assignment. Furthermore, there is *no* way to

achieve $A[z] = A[z'] + s$ **varying** z,

since the value $A[z'] + s$ might not appear in A at all.

So we must look for another alternative. Since $A[u] = A[\lfloor u \rfloor]$, our goal

achieve $A[z] = b$ **varying** z

is equivalent to

achieve $A[\lfloor z \rfloor] = b$ **varying** z

At this point, we would like to extract z from the expression $A[\lfloor z \rfloor]$, as it appears by itself in the output invariants of the given program. To facilitate this, we need to make temporary use of an inverse of the array-indexing function.[10] The function $pos(U, u)$ gives the position of the (rightmost) occurrence of the element u in the array U; it is defined by the following three facts:

---

[10]The inverse function serves only as a formal mechanism for expressing transformations. It will be eliminated from the final version of the program. Cf. [Kirchner-KirchnerJouannaud81].

> **fact** $pos(U,u) \in \mathbf{Z}$ **when** $u \in \{U\}$
> **fact** $U[pos(U,u)]=u$ **when** $u \in \{U\}$
> **fact** $pos(U,u) > v-1$ **when** $U[v]=u$,

where $\{U\}$ denotes the multiset of elements in U. Instantiating the second fact, with b in place of u and A in place of U, yields $A[pos(A,b)]=b$ (it is given that $b \in \{A\}$). Thus, in order to

> **achieve** $A[\lfloor z \rfloor]=b$ **varying** z,

it suffices to

> **achieve** $pos(A,b)=\lfloor z \rfloor$ **varying** z.

Applying now the definition of $\lfloor u \rfloor$,

> **fact** $v=\lfloor u \rfloor$ **is** $v \leq u$, $u < v+1$, $v \in \mathbf{Z}$,

we obtain the conjunctive goal

> **achieve** $pos(A,b) \leq z$, $z < pos(A,b)+1$, $pos(A,b) \in \mathbf{Z}$ **varying** z.

Since the third conjunct $pos(A,b) \in \mathbf{Z}$ follows from $b \in \{A\}$, we are left with the goal

> **achieve** $pos(A,b) \leq z$, $z < pos(A,b)+1$ **varying** z.

The current goal is still not readily comparable with the output specification of the real square-root program,

> **assert** $r \leq \sqrt{a}$, $\sqrt{a} < r+e$.

Whereas for the array-search program the output variable z appears on the right-hand side of the $\leq$ relation and on the left-hand side of the $<$ relation, for the square-root program the sides are reversed. One possible solution is to transform the predicates $\leq$ and $<$. To get

$$r \leq \sqrt{a} \quad \Rightarrow \quad z \geq pos(A,b),$$

we may apply the transformations

$$\begin{array}{ccc} \leq & \Rightarrow & \geq \\ \sqrt{a} & \Rightarrow & pos(A,b). \end{array}$$

To obtain the second transformation, we let

$$a \quad \Rightarrow \quad pos(A,b)^2.$$

Applying these transformations to the conjunct $\sqrt{a} < r + e$ leaves

$$pos(A,b) < z + e \quad \Rightarrow \quad pos(A,b) + 1 > z.$$

Transposing to isolate $pos(A,b)$ on both inequalities, gives

$$pos(A,b) < z + e \quad \Rightarrow \quad pos(A,b) > z - 1,$$

so we add the transformations $< \Rightarrow >$ and $e \Rightarrow -1$. All together we have[11]

$$
\begin{aligned}
\leq &\quad \Rightarrow \quad \geq \\
a &\quad \Rightarrow \quad pos(A,b)^2 \\
< &\quad \Rightarrow \quad > \\
e &\quad \Rightarrow \quad -1.
\end{aligned}
$$

Applying these four transformations to the given square-root program, and simplifying, yields the program

```
(z,s) := (1, pos(A,b)² - 1)
loop suggest z ≥ pos(A,b), pos(A,b) > z + s
     until s ≥ -1
     s := s/2
     if z + s ≥ pos(A,b) then z := z + s fi
     repeat
suggest z ≥ pos(A,b), pos(A,b) > z - 1 .
```

(The transformed invariants are only suggestions, since the constants $\leq$ and $<$ are being transformed.) Before we try to eliminate the nonprimitive function pos from the transformed program, we attempt to verify the correctness of the program as is. The suggested loop invariants

$$\textbf{suggest } z \geq pos(A,b), \ pos(A,b) > z + s$$

along with the exit condition

---

[11]There is another way to change the specifications so as to highlight a possible analogy. Just as $u \leq v$ is equivalent to $u < v + 1$ for integers u and v, for real numbers one can replace an expression of the form $u \leq v$ with one of form $u < v + \epsilon$, where ϵ represents an arbitrarily small real number. Comparing the new specifications would suggest the transformations $a \Rightarrow pos(A,b)^2$, $z \Rightarrow z - 1 + \epsilon$, and $e \Rightarrow 1$. The ϵ's could then be eliminated from the resulting program.

until $s \geq -1$

clearly imply the desired output invariant

suggest $pos(A,b) \leq z$, $z < pos(A,b) + 1$.

Furthermore, the loop-body path preserves the loop invariants for both cases of the conditional.

The problem is with the verification condition for the initialization path: the assignment $(z,s) := (1, pos(A,b)^2 - 1)$ does not initialize the loop invariants. Since the transformed initialization does not work, we must replace it with an unachieved subgoal

assert $u \leq v \supset A[u] \leq A[v]$, $A[u] \in \mathbf{Z}$, $b \in \{A[1:n]\}$
achieve $z \geq pos(A,b)$, $pos(A,b) > z + s$ **varying** z,s.

The purpose of this goal is to set the variables z and s so that the loop invariants $z \geq pos(A,b)$ and $pos(A,b) > z + s$ hold when the loop is entered for the first time. Since we are given that b appears within the segment $A[1:n]$, we can achieve the relation $z \geq pos(A,b)$ by letting $z = n$. Now we can achieve $pos(A,b) > z + s$ by insisting that $z + s = 0$, for which we initialize s to $-z = -n$.

Replacing the initialization requires rechecking the verification condition for termination. The termination condition is

$$(\exists \xi \in \mathbf{N}) \ -\frac{n}{2^\xi} \geq -1,$$

i.e. by repeatedly halving s, which has the initial value $-n$, the exit test $s \geq -1$ must at some point become true.[12] This is indeed the case. Thus, all the verification conditions hold and the transformed program is correct.

Finally, the conditional test $z + s \geq pos(A,b)$, containing the nonprimitive function pos, may be replaced by $A[z + s + 1] > b$. That the two tests are equivalent, may be deduced from the input specification $u \leq v \supset A[u] \leq A[v]$ and the definition of pos. Our program now looks like this:

---

[12]This is the "exit method" of proving termination; see [KatzManna75b].

$$(z,s) := (n,-n)$$

loop L_1: **assert** $A[z+s+1] \leq b,\ b < A[z+1]$
  until $s \geq -1$
  $s := s/2$
  if $A[z+s+1] > b$ **then** $z := z+s$ **fi**
  repeat
E_1: **assert** $A[z]=b$.

3.3.3.2. Transformation

The above array-search program may be given a more conventional appearance by replacing $z+s$ (the lower bound of the search) with a new variable y. For this purpose, we use another global transformation:

$$s \quad \Rightarrow \quad y-z.$$

Note that transforming a variable that does not appear in the program specification, such as s, cannot affect what the program does, only how it does it. Since the right-hand side of the transformation contains a variable already appearing in the program, viz. z, the application of the transformation is a bit tricky: whenever there is an assignment to z, even if s is not changed, one must consider what happens to the transformed value of s. Thus, the **then**-branch assignment $z := z+s$ must be considered as though it were $(z,s) := (z+s,s)$, which transforms into $(z,y-z) := (z+(y-z),y-z)$. Applying the transformation, and using the appropriate **achieve** statements in place of illegal assignments, we get

  achieve $z=n,\ y-z=-n$ **varying** z,y
  loop L_1: **assert** $A[y+1] \leq b,\ b < A[z+1]$
    until $z-y \leq 1$
    achieve $y-z=(y'-z)/2$ **varying** y
    if $A[y+1] > b$ **then**
      achieve $z=z'+(y'-z'),\ y-z=y'-z'$
        varying z,y **fi**
  repeat,

where y' and z' denote the values of the variables y and z, respectively, prior to the **achieve** statement within which they appear.

The initialization subgoal

achieve $z=n$, $y-z=-n$ **varying** z,y

yields the assignment

$$(y,z) := (0,n).$$

Isolating y on one side of the equality in the subgoal

achieve $y-z=(y'-z)/2$ **varying** y,

yields

$$y := (z+y)/2.$$

Similarly, the **then**-branch subgoal

achieve $z=z'+(y'-z')$, $y-z=y'-z'$ **varying** z,y

yields the assignment

$$(y,z) := (2{\cdot}y-z,y).$$

The program, thus far, is

$$(y,z) := (0,n)$$
$$\textbf{loop } L_1: \textbf{ assert } A[y+1] \leq b,\ b < A[z+1]$$
$$\textbf{until } z-y \leq 1$$
$$y := (z+y)/2$$
$$\textbf{if } A[y+1] > b \textbf{ then } (y,z) := (2{\cdot}y-z,y) \textbf{ fi}$$
$$\textbf{repeat}.$$

There are still a few more changes that may be made: "distributing" the loop-body assignment over the two branches of the conditional statement, gives

$$\textbf{if } A[(z+y)/2+1] > b \quad \textbf{then} \quad (y,z) := (y,(z+y)/2)$$
$$\textbf{else} \quad y := (z+y)/2 \qquad \textbf{fi}.$$

Now, by eliminating the superfluous assignment $y := y$ and introducing a temporary variable t to contain the value $(z+y)/2$, we get our final version of the array-search program:

R_1: **begin comment** *array-search program*
  B_1: **assert** $u \leq v \supset A[u] \leq A[v]$, $A[u] \in \mathbf{Z}$, $b \in \{A[1:n]\}$
  $(y,z) := (0,n)$
  loop L_1: **assert** $A[y+1] \leq b$, $b < A[z+1]$
      until $z - y \leq 1$
      $t := (z+y)/2$
      if $A[t+1] > b$ **then** $z := t$ **else** $y := t$ **fi**
      repeat
  E_1: **assert** $A[z]=b$
  end

3.4. Discussion

Given an old program and a specification for a new program, the task of modifying the given program to satisfy the desired specification can be subdivided into eight steps. These steps fall into three phases: a pre-modification phase that sets up the given program and its specification; the modification phase itself as presented in the previous sections; and a post-modification phase that uses additional techniques to complete the task.

3.4.1. Pre-modification Phase

1. *Annotate the given program.* If necessary annotate the given program with invariant assertions, so that modification methods can be applied. In particular, determine the relation between input and output variables achieved by the program. Annotation is essential for debugging, since it is not even known what the incorrect program actually does. We discuss annotation in Chapter 6.

2. *Rephrase the specifications to bring out their similarity.* The specifications of the given and desired programs may be given in a form that obscures any analogy. Thus, one would like to express the specifications in some equivalent form which makes their similarity more pronounced. This in general may be a very difficult task.

3.4.2. Modification Phase

3. *Discover an analogy between the specifications.* This analogy consists of a set of transformations that yield the desired specification when applied to the specification of the given program.

4. *Check the validity of the proposed modification.* For those types of transformations that do not necessarily preserve the correctness of the program, the verification conditions are checked. Sometimes the analogy must be extended with additional transformations for the conditions to hold. The next chapter describes this step in greater detail.

5. *Apply the transformations to the program.* The transformations found in the previous two steps are applied to the given program. The previous step guarantees that the modified program will meet its specification.

6. *Rewrite any unexecutable statements.* If a variable being assigned to has been transformed into an expression, then the assignment statement must be replaced. Any nonprimitive expressions introduced into the program by the transformations must be reexpressed in terms of primitives.

3.4.3. Post-modification Phase

7. *Synthesize new segments.* If necessary, write new program segments
    to replace parts of the program that cannot be modified. Also
    extend the modified program for any unachieved parts of the desired
    specification, by integrating code into the program. Synthesis and
    extension are discussed in Chapter 5.

8. *Optimize the transformed program.* It may be possible to optimize
    the new program by taking advantage of properties it has that may
    not have held for the old program. Domain-specific transformations
    can also be applied.

Chapter 4

Program Abstraction and Instantiation

*The data with which the sciences start out are concrete,
whereas the objectives they strive for are abstract.*

*—Saadia ben Joseph al-Fayyumi Gaon (The book of beliefs
and opinions)*

*Each problem that I solved became a rule which served
afterwards to solve other problems.*

*—René Descartes (Discourse on the method of rightly
conducting the reason and seeking truth in the field of
science)*

*As soon as men recognize some similarity between two
things, it is their custom to ascribe to each of them, even in
those respects in which they are different, what they know to
be true of the other.*

—René Descartes (Rules for the direction of the mind)

4.1. Introduction

When confronted with a new task, a human often recognizes some measure of resemblance between it and another, previously accomplished, task. Rather than "reinvent the wheel," he is prone to conserve effort by adapting the known solution of the old problem to the problem now at hand. Then, after having solved several related problems, he might come to formulate a general paradigm for solving that type of problem by highlighting the shared aspects of the individual instances and suppressing their inconsequential or idiosyncratic particulars. This process of formulating a general scheme from concrete instances is termed *abstraction;* that of applying an abstract scheme to a particular problem is termed *instantiation.*

Like modification and debugging, abstraction and instantiation are typical phases in the evolutionary cycle of many programs. The more experience a programmer has had, the more programming methods he is likely to have assimilated, and the more judiciously he can apply them to new problems. Thus, after constructing a number of similar programs, a programmer is apt to formulate for himself—and perhaps for others as well—an abstract notion of the underlying principle and reuse it in solving related problems. *Program schemata* are a convenient form for remembering such programming knowledge. A schema may embody basic programming techniques or specialized strategies for solving some class of problems; its specification is stated in terms of abstract predicate, function, and constant symbols. For example, the binary-search technique can be expressed as a schema that can be instantiated to compute quotients or roots.

To formalize this aspect of programming, we develop methods for abstracting a given set of cognate programs into a program schema and for instantiating a schema to satisfy a given concrete specification. To date there has been a limited amount of research on program abstraction. The STRIPS system ([FikesHartNilsson72]) generalized the loop-free robot plans that it generated; the HACKER system ([Sussman75]) "subroutinized" and generalized the "blocks-world" plans it synthesized, executing the plan to determine which program constants could be abstracted. In a similar vein, [Plaisted80] demonstrates how abstractions may be used, with great effectivity, for theorem proving. [DershowitzManna75] suggested using the proof of correctness of a program to guide the abstraction process.

A number of researchers have described the use of program schemata. [Dijkstra72] maintains that theorems about schemata are unconsciously invoked by programmers; [ConwayGries75] and [Wirth76] illustrate the use of basic schemata in the systematic development of programs. [Gerhart75b], [GerhartYelowitz76b], [YelowitzDuncan77], [Muralidharan82], and others have advocated and illustrated the use of schemata as a powerful programming tool. Along similar lines, [RichWaters81] suggests that a simplified abstract model of a program should be filled in and then debugged, and [Brotsky81] recommends that programs be analyzed in terms of "cliche's." Related notions are "programming plans," as described, for example, in [Waters82] or [Barstow79], and "generic units," as used, for example, in [Ada81]. Plans are designed to include more semantic information than schemata; generic facilities, less. Some other examples of the use of abstract or simplistic algorithms as first steps in the development of programs are [Darlington78], [Deussen79], [DuncanYelowitz79], [LeeRoeverGerhart79], and [BackManillaRaiha83]. An approach to the specification and verification of program schemata is given in [Misra78].

We suggest the formulation of analogies as a basic tool in program abstraction, similar to their use for modification. First, an analogy is sought between the output specifications of the given programs. This yields an abstract specification that may be instantiated to any of the given concrete specifications. The analogy may then used as a basis for transforming the existing programs into an abstract schema that represents the embedded technique. Our methodology is applicable to all programming styles: iterative or recursive, declarative or applicative. We use invariant assertions and correctness proofs to help extend and complete the analogy. The more information available regarding the underlying rationale of a program, the more directed the abstraction process can be.

A schema, derived in this manner, is usually not applicable to all possible instantiations of its specifications. For that reason, a schema is, in general, accompanied by an input specification containing conditions that must be satisfied by the instantiation in order to guarantee correctness. These *preconditions* are derived from a correctness proof of the schema. The abstract specification of the schema may then be compared with a given concrete specification of a new problem. By formulating an analogy between the abstract and concrete specifications, an instantiation is found that yields a concrete program when applied to the abstract schema. If the instantiation satisfies the preconditions, then the correctness of the new program is guaranteed. If not, analysis of the unsatisfied conditions may suggest modifications that will lead to a correct program.

Deriving a general correctness-preserving program transformation from a set of examples is similar to deriving a schema from a set of concrete programs; we show how the same approach works for both tasks. The importance of program transformations as a tool for program development has been pointed out by many, including [Knuth74], [DarlingtonBurstall76], and [Wegbreit76] (for a more pessimistic view, see [Dijkstra78]). [Gerhart75b] and others have recommended the hand-compilation of a handbook of program schemata. Such a collection of schemata, along with a catalog of correctness-preserving program transformations, could serve as part of an interactive program-development system. Some schemata that have appeared in the programming literature appear in Appendix 2; collections of transformations include [AllenCocke72], [Bauer76], [Standish,*etal.*76], [Loveman77], [Schwartz77], [Arsac79], [KieburtzShultis81], and [Wadler81].

In the next section we give an overview of the method. Section 4.3 presents several representative illustrations of the approach and is followed by a discussion.

4.2. Overview

Abstracting a set of programs begins with finding similarities and accentuating differences. Based on that step, a set of transformations is applied to the programs. The result is a schema that instantiates to the given programs. For example, one of the simplest programming schemata is the following:[1]

S_4: **begin** *linear–search schema*
$\quad z := 0$
$\quad$ **loop until** $\tau(z)$
$\qquad z := z + 1$
$\qquad$ **repeat**
$\quad E_4$: **assert** $z = \min\limits_{n \in N} \tau(n)$
$\quad$ **end**

Given a predicate τ, this schema searches the nonnegative integers linearly for the first one to satisfy τ. In general, to determine exactly under which conditions a schema is applicable, its correctness proof must be analyzed. The above schema, for example, will only terminate if there is some nonnegative integer for which τ is true. In this section, we give an outline of the abstraction process. We also describe the similar steps necessary to instantiate a schema when confronted with a new problem, and to derive a general "program transformation" from a set of examples.

4.2.1. Basic Method: Analogy

Two objects can be compared on many levels: external appearance, outward performance, inner workings. Generally, the "external appearance" of a program, i.e. the code, can undergo dramatic changes without affecting the underlying algorithm. At the opposite end of the spectrum, input/output specifications, defining the "outward performance" of a program, but not how it accomplishes what it does, can be identical for very disparate programs. Between the two extremes, comments about the program's "inner workings," its correctness and efficiency, are perhaps a better guide when looking for similarities between programs. Therefore, it makes sense to begin by formulating an analogy between program specifications, and then extending that analogy by examining how the different programs achieve their analogous desiderata. The more

---

[1][Dijkstra72]

that is known about the "rhyme and reason" of a program, the better the chances of being able to profit from comparing it with other programs.

When the specification for, and documentation of, a program are given formally, then we can use formal means to compare them, as we did in the previous chapter when we attempted to modify programs to perform new tasks. For example, if one program is known to end with the condition $z \leq A[0]$ holding, and a second program halts with $B[p] \geq B[k]$, then we look for an analogy

$$z \leq A[0] \quad \Longleftrightarrow \quad B[p] \geq B[k]$$

between the two. In this case, we can say that if $\leq$ in the first condition corresponds with $\geq$ in the second, then z and $A[0]$ in the first ought to correspond, in the second, with $B[p]$ and $B[k]$, respectively. Going a step further, we can say that where the former condition has 0, $\leq$, A, and z, the latter has k, $\geq$, B, and $B[p]$, respectively. This, more detailed, analogy is denoted

$$
\begin{array}{ccc}
0 & \Longleftrightarrow & k \\
\leq & \Longleftrightarrow & \geq \\
A & \Longleftrightarrow & B \\
z & \Longleftrightarrow & B[p].
\end{array}
$$

In general, there are several possible ways in which an analogy

$$f(q,r) \quad \Longleftrightarrow \quad h$$

between an expression of the form $f(q,r)$ and another expression h can be refined:

● If h is of the form $g(s,t)$, that is, if we are comparing

$$f(q,r) \quad \Longleftrightarrow \quad g(s,t),$$

then the *imitating* mapping

$$
\begin{array}{ccc}
f & \Longleftrightarrow & g \\
q & \Longleftrightarrow & s \\
r & \Longleftrightarrow & t
\end{array}
$$

suggests itself. These three correspondences may be broken down further, as long the mappings remain consistent with each other.

- If f has an inverse f^- in its first argument, i.e. $f(f^-(x,y),y)=x$ (for all x and y), then an *inverting* mapping

$$q \iff f^-(h,r)$$

can be used.

- If f has an identity element f^0 in its first argument, i.e. $f(f^0,v)=v$ (for all v), then a *collapsing* mapping

$$\begin{array}{ccc} q & \iff & f^0 \\ r & \iff & h \end{array}$$

is possible.

- Another possibility would be the *projecting* mapping

$$\begin{array}{ccc} f(u,v) & \iff & u \\ r & \iff & h, \end{array}$$

i.e. f maps to a function that projects its first argument.[2]

For example, in comparing

$$z \leq A[0] \iff B[p] \geq B[k],$$

the imitating mapping

$$\begin{array}{ccc} 0 & \iff & k \\ z \leq A[u] & \iff & B[p] \geq B[u] \end{array}$$

is first used. To compare $z \leq A[u]$ with $B[p] \geq B[u]$, another imitating mapping is used, viz.

$$\begin{array}{ccc} \leq & \iff & \geq \\ A & \iff & B \\ z & \iff & B[p]. \end{array}$$

As another example, comparing the two expressions

---

[2]Similar mappings work for other than the first argument. For a discussion of a second-order pattern-matcher that uses imitating and projecting mappings and its application to program manipulation see [HuetLang78].

$$c/d \quad \Longleftrightarrow \quad \sqrt{a} \,,$$

suggests numerous possibilities, including: the imitating mapping

$$u/v \quad \Longleftrightarrow \quad \sqrt{v}$$
$$d \quad \Longleftrightarrow \quad a \,,$$

the collapsing mapping

$$d \quad \Longleftrightarrow \quad 1$$
$$c \quad \Longleftrightarrow \quad \sqrt{a} \,,$$

and the projecting mapping

$$u/v \quad \Longleftrightarrow \quad u$$
$$c \quad \Longleftrightarrow \quad \sqrt{a} \,.$$

Note that the projecting mapping gives only partial information: it indicates how to get from c/d to $\sqrt{a}$; it does not, however, indicate how to go in the other direction, since the right-hand side u of the analogy can match *anything*.

Given an analogy such as

$$0 \quad \Longleftrightarrow \quad k$$
$$\leq \quad \Longleftrightarrow \quad \geq$$
$$A \quad \Longleftrightarrow \quad B$$
$$z \quad \Longleftrightarrow \quad B[p],$$

we can define abstract entities for corresponding parts. Each pair in the analogy is denoted by an abstract variable of the same kind: the scalar constant 0 and variable k may be replaced by an abstract variable κ; the predicate constants $\leq$ and $\geq$ are abstracted to a predicate variable α; the array variables A and B generalize to a function variable Δ; and the variable z and variable expression $B[p]$ generalize to a variable μ.[3] Applying the transformations

---

[3]In this chapter, we will use Greek letters to distinguish *abstract* entities.

$$
\begin{array}{ccc}
0 & \Rightarrow & \kappa \\
\leq & \Rightarrow & \alpha \\
A & \Rightarrow & \Delta \\
z & \Rightarrow & \mu
\end{array}
$$

to the predicate $z \leq A[0]$ yields the abstract condition $\alpha(\mu, \Delta[\kappa])$. Such a set of transformations is called an *abstraction mapping*. Similarly, the abstraction mapping

$$
\begin{array}{ccc}
k & \Rightarrow & \kappa \\
\geq & \Rightarrow & \alpha \\
B & \Rightarrow & \Delta \\
B[p] & \Rightarrow & \mu,
\end{array}
$$

when applied to the condition $B[p] \geq B[k]$, yields the same abstract condition $\alpha(\mu, \Delta[\kappa])$.[4]

The first step in the abstraction process is the attempt to find a detailed analogy between the specifications of the given programs. Then, the tentative abstraction mapping may be applied to the corresponding program. The result is a program schema containing abstract symbols in place of the concrete entities that were in the original program. At other stages in the abstraction process, additional correspondences between programs may be found that lead to further abstractions.

4.2.2. Correctness Considerations

Global transformations of variables (such as $A \Rightarrow \Delta$ and $z \Rightarrow \mu$ in the above example), when applied to all occurrences of those variables in an annotated program, of necessity result in a correct schema, just as was the case with the global modifications described in the previous chapter. But when, as usually happens, an abstraction mapping also involves transformations of constant symbols (e.g. 0 and $\leq$ in the above example), the resultant schema will not necessarily be correct. That is

---

[4] We should be careful and specify the order in which the individual mappings are applied. If $B \Rightarrow \Delta$ is applied first, then we will have to transform $\Delta[p] \Rightarrow \mu$ instead of $B[p] \Rightarrow \mu$.

because the correctness of the original programs presumably depends on specific properties of those constants. For example, were there a superfluous assignment $i := i + 0$ in a program, then an abstraction of 0 would overzealously change that assignment, as well as any relevant occurrences. Or, were some function symbol to appear in the specification, but not explicitly in the code, then transforming that symbol would be ineffectual.

When an analysis of the abstracted program reveals that it is not necessarily correct, four remedies are available: localized transformations, extended analogies, additional preconditions, and inserted subgoals. The first thing to do is to try to identify exactly which occurrences of the various symbols in the given programs are analogous. Then, rather than apply an abstraction mapping blindly to all occurrences, the transformations may be localized to the relevant ones only. For this to be possible, we need information about how the original programs were constructed, explaining which parts of the specifications are related to which parts of the programs. Ideally, all programs would come with detailed comments showing each stage in the refinement of the specifications step-by-step towards the final version of the program (as we do for the programs we synthesize in Chapter 5). In the absence of such, one can attempt to derive invariants from the code (as we do in Chapter 6), prove the program's correctness, and then extract the necessary data from the proof.

Localizing transformations will avoid overzealousness, but it may also be the case that the analogy between the specifications gives less than the "whole story." In that case, it may be possible to extend the analogy between the programs by examining them in more detail. When programs are solving similar problems in similar manners, it is likely that their major components also have similar goals. Therefore, it should be possible to continue comparing the programs, level by level, down to the "gory" details of their code, until the analogy breaks down at fine points over which the programs really differ. With each extension of the analogy, abstract entities are added to the developing schema.

Usually, even after finding a complete analogy, it will not be possible to prove the schema correct. The reason is that the correctness of the original programs was based on certain properties of those symbols that were abstracted. For an instance of the schema to work properly, those properties must also hold for the instantiation. What we need to do, is to accompany each schema with *preconditions* that specify exactly under which circumstances an instantiation will yield a correct program. These conditions are derived from an analysis of the conditions necessary for a proof a correctness of the schema; they should, of course, hold for the given set of programs. (Verifying the correctness of schemata is no different than verifying programs.) Some amount of heuristic reasoning could also be used at this point to generalize them.

Sometimes, there may be no set of preconditions that ensures correctness of the complete schema. Nevertheless, those parts that are correct may be useful, while the others can be replaced with unachieved subgoals. It may turn out, for example, that the abstracted initialization of a loop does not work as required, in which case the initialization part might be replaced by a subgoal stating—in terms of the schema's abstract entities—that the variables appearing in the loop invariants must be set so that the invariants hold when the loop is entered.

4.2.3. Instantiation

Once we have a collection of program schemata, we would like to be able to exploit them in the solution of new problems. This is accomplished by looking for a schema whose abstract specification can be compared with the concrete specification of the desired goal. The comparison suggests an *instantiation mapping,* that, when applied to the schema, yields a concrete program for the given problem. If the instantiated preconditions hold in the new problem domain, the instantiated schema is a correct program.

As an illustration, consider again the linear search schema

S_4: **begin** *linear–search schema*
B_4: **assert** $(\exists\, u \in N)\; \tau(u)$
$z := 0$
loop   until $\tau(z)$
        $z := z + 1$
        repeat
E_4: **assert** $z = \underset{n \in N}{min}\; \tau(n)$
end

The precondition

$$B_4: \textbf{assert}\; (\exists\, u \in N)\; \tau(u)$$

requires that τ be true for *some* nonnegative integer, which ensures that the instantiated program will eventually terminate. Now, suppose we wish to find the "integer square-root" of an input value a. That is, our goal is to construct a program that sets the value of a variable z to $\lfloor \sqrt{a} \rfloor$, where $a \in N$ and $\lfloor u \rfloor$ is the largest integer less than or equal to u:

P_4: **begin comment** *integer square–root specification*
assert $a \in N$
achieve $z = \lfloor \sqrt{a} \rfloor$ **varying** z
end

If z is to be the largest integer not greater than $\sqrt{a}$, then $z + 1$ must be the smallest integer greater than $\sqrt{a}$. It follows that our goal will be attained if we can

$$\textbf{achieve}\; z = \underset{n \in N}{min}\; (n + 1 > \sqrt{a}).$$

This version of the integer-square root specification matches the abstract output specification

$$\textbf{assert}\; z = \underset{n \in N}{min}\; \tau(n)$$

of our schema. Applying the appropriate instantiation mapping

$$\tau(u) \quad \Rightarrow \quad u + 1 > \sqrt{a}$$

to the schema, replaces the exit test

$$\textbf{until } \tau(z)$$

with

$$\textbf{until } z+1 > \sqrt{a}$$

After applying an instantiation mapping to a schema, one must make sure that all the statements are executable. If we assume, as is reasonable, that the square-root function itself is not a primitive operation for this problem, then the above test is not executable. Fortunately, we can instead test

$$\textbf{until } (z+1)^2 > a,$$

providing us with the following integer square-root program:

P_4: **begin** *simple integer square-root program*
  B_4: **assert** $a \in \mathbf{N}$
  $z := 0$
  loop until $(z+1)^2 > a$
    $z := z+1$
    repeat
  E_4: **assert** $z = \lfloor \sqrt{a} \rfloor$
  end

This program is correct (though not particularly inspiring), since the instantiation of τ satisfies the precondition

$$(\exists\, u \in \mathbf{N})\ u+1 > \sqrt{a}.$$

4.2.4. Special Case: Program Transformations

Designing and applying program transformations also fall into the abstraction/instantiation paradigm. The same methodology can be used for this purpose as for designing and applying program schemata. For example, the recursive program

> $P_6'(m,n)$: **begin comment** *recursive factorial program*
>   **if** $n=0$ **then** $m := 1$
>   **else** $P_6'(m,n-1)$
> $m := m \cdot n$ **fi**
> E_6': **assert** $m=n!$
>   **end**

for computing factorials, can be transformed into the iterative program

> P_6: **begin comment** *iterative factorial program*
> $(m,y) := (1,0)$
>   **loop**   **until** $y=n$
> $(m,y) := (m \cdot (y+1), y+1)$
>   **repeat**
> E_6: **assert** $m=n!$
>   **end**

In a similar manner, recursion may be eliminated from the program

> $Q_6'(k,s)$: **begin comment** *recursive summation program*
>   **if** $k=l+1$ **then** $s := 0$
>   **else** $Q_6'(k+1,s)$
> $s := s+f(k)$ **fi**
> E_6': **assert** $s=\sum_{i=k}^{l} f(i)$
>   **end**

for summing a function f over the range $[k:l]$.

    What we would like to do is to use these programs to design a program transformation schema of the form

> R: *recursive schema*
> S: *iterative schema*

meaning that a program R may be replaced by the corresponding program S (if the input specification of S is satisfied). To that end, we begin with a comparison of the output specifications of the two programs:

$$m=n! \quad \Longleftrightarrow \quad s=\sum_{i=k}^{l} f(i).$$

One possible analogy between them is

$$
\begin{array}{ccc}
m & \Longleftrightarrow & s \\
n & \Longleftrightarrow & k \\
u! & \Longleftrightarrow & \sum_{i=u}^{l} f(i).
\end{array}
$$

This analogy does not supply much information; it can, however, be extended by comparing the two recursive programs P_6' and Q_6'. The exit tests $n=0$ and $k=l+1$ add

$$
0 \quad \Longleftrightarrow \quad l+1
$$

to the analogy; the base cases $m := 1$ and $s := 0$ add

$$
1 \quad \Longleftrightarrow \quad 0;
$$

the recursive calls $P_6'(m,n-1)$ and $Q_6'(k+1,s)$ add

$$
\begin{array}{ccc}
P_6'(u,v) & \Longleftrightarrow & Q_6'(v,u) \\
u-1 & \Longleftrightarrow & u+1,
\end{array}
$$

since m has already been matched up with s and n with k; and, lastly, the **else**-branch assignments $m := m \cdot n$ and $s := s + f(k)$ add

$$
u \cdot v \quad \Longleftrightarrow \quad u + f(v).
$$

The above analogy is still insufficient, however, and it would take an analysis of the iterative programs to complete it with

$$
u+1 \quad \Longleftrightarrow \quad u-1.
$$

Defining abstract entities for each pair in the analogy, and applying the abstraction mapping

$$
\begin{array}{ccc}
m & \Rightarrow & \mu \\
n & \Rightarrow & \nu \\
u! & \Rightarrow & \phi(u) \\
0 & \Rightarrow & \theta \\
1 & \Rightarrow & \omega \\
P_6'(u,v) & \Rightarrow & R_6(u,v) \\
u-1 & \Rightarrow & \delta(u) \\
u \cdot v & \Rightarrow & \psi(u,v) \\
u+1 & \Rightarrow & \gamma(u)
\end{array}
$$

to the recursive program P_6' gives

$R_6(\mu,\nu)$: **begin comment** *recursion schema*
  if $\nu=\theta$ **then** $\mu := \omega$
          else $R_6(\mu,\delta(\nu))$
                $\mu := \psi(\mu,\nu)$ **fi**
  E_6: **assert** $\mu=\phi(\nu)$
  end

The same mapping, applied to the iterative version P_6, produces the abstract iterative program

S_6: **begin comment** *invertible recursion schema*
  $(\mu,y) := (\omega,\theta)$
  loop **until** $y=\nu$
      $(\mu,y) := (\psi(\mu,\gamma(y)),\gamma(y))$
      repeat
  E_6: **assert** $\mu=\phi(\nu)$
  end

An analysis of the verification conditions of this schema would reveal that the program-transformation

R_6: *recursion schema*
S_6: *invertible recursion schema*

is correct under the assumption that the function γ is the inverse of δ. Any recursive program of the same form as R_6 can be transformed into an iterative program of the form S_6, if the inverse function γ is known.

Many recursion removal transformations have appeared in the literature, including [McCarthy62], [Cooper66], [PatersonHewitt70], [Walker-Strong73], [Knuth74], [DarlingtonBurstall76], [Bird77], [Rohl77], [Wossner,*etal.*78], and [Hikita79]. How to demonstrate the correctness of program transformations is considered in [Gerhart75a], [HuetLang78], [BroyKriegBruckner80], and others.

4.3. Examples

This section contains three detailed examples of program abstraction and instantiation. The first example is relatively straightforward. An analogy between the specifications of two essentially similar programs is used to transform one of them into an abstract schema. An analysis of the schema's verification conditions suggests an additional transformation and supplies preconditions that guarantee the correctness of an instantiation. The second example illustrates how detailed comments on the workings of the given programs, and the reasonings that went into their development, can aid the abstraction process. The last example shows how the same methodology may be used for the development and application of program transformations.

4.3.1. Extremum

Consider the following program for finding the position p of a maximal element in an array segment $B[b:c]$:

```
Q₅: begin comment maximum–position program
  B₅: assert b ∈ Z, c ∈ Z, b ≤ c
  (p,j) := (c,c)
  loop  L₅: assert B[p] ≥ B[j:c], p ∈ Z, j ∈ Z
        while j > b
        j := j - 1
        if B[j] > B[p] then p := j fi
        repeat
  E₅: assert B[p] ≥ B[b:c]
  end
```

Its output specification is

$$E_5: \textbf{assert } B[p] \geq B[b:c],$$

meaning that $B[p]$ is greater or equal to each element $B[u]$, where u is an integer in the range b to c, inclusive. Its input specification,

$$B_5: \textbf{assert } b \in \mathbf{Z}, \, c \in \mathbf{Z}, \, b \leq c$$

(where $\mathbf{Z}$ is the set of all integers), requires that the two input variables

b and c be integers and that b not be greater than c so that the array segment $B[b:c]$ is nonempty. The loop decrements j as long as the condition

$$\textbf{while } j > b$$

still holds. This program is similar to the following one:

P_5: **begin comment** *minimum–value program*
  B_5: **assert** $n \in \mathbf{N}$
  $(z,i) := (A[0],0)$
  loop L_5: **assert** $z \leq A[0:i]$, $i \in \mathbf{N}$
      until $i = n$
      $i := i + 1$
      if $A[i] < z$ **then** $z := A[i]$ **fi**
      repeat
  E_5: **assert** $z \leq A[0:n]$
  end

The latter's output specification

$$E_5: \textbf{assert } z \leq A[0:n]$$

requires that the value of z not be greater than any value appearing in the array segment $A[0:n]$.

Both programs search for an extremum element in an array segment. Our task is to extract an abstract version of these two programs that captures the essence of the technique used, but that is not specific to either problem. The resultant schema could then be used as a model of such searches for the solution of future problems.

The first step in abstracting these two programs P_5 and Q_5 is to find an analogy between their respective output specifications $z \leq A[0:n]$ and $B[p] \geq B[b:c]$. A straightforward analogy is that where the specification of P_5 has 0, n, $\leq$, z, and A, the specification of Q_5 has b, c, $\geq$, $B[p]$, and B, respectively:

$$
\begin{array}{ccc}
0 & \Longleftrightarrow & b \\
n & \Longleftrightarrow & c \\
\leq & \Longleftrightarrow & \geq \\
z & \Longleftrightarrow & B[p] \\
A & \Longleftrightarrow & B.
\end{array}
$$

Now abstract entities are substituted for analogous parts. For the scalar constant 0 in P_5 and the corresponding scalar input variable b in Q_5, we use an abstract input variable κ. The corresponding input variables n and c may be replaced by λ. We replace the predicate constants $\leq$ and $\geq$ with an abstract predicate α. The input arrays A and B generalize to an arbitrary function Δ, while the output variable z and the variable expression $B[p]$ are generalized to the program variable μ. The corresponding abstraction mappings are

$$
\begin{array}{ccccc}
0 & \Rightarrow & \kappa & \Leftarrow & b \\
n & \Rightarrow & \lambda & \Leftarrow & c \\
\leq & \Rightarrow & \alpha & \Leftarrow & \geq \\
z & \Rightarrow & \varsigma & \Leftarrow & B[p] \\
A & \Rightarrow & \Delta & \Leftarrow & B,
\end{array}
$$

where the leftmost column has symbols from P_5, the rightmost column has symbols from Q_5, and the center column has their abstract counterparts. To transform $B[p] \Rightarrow \varsigma$ for Q_5, we can transform $p \Rightarrow pos(\Delta, \varsigma)$, where (as in Section 3.3.3.1) pos denotes the (formal) inverse of the array access function, i.e. pos is defined so that $U[pos(U,v)]=v$ for any array U and value v. Applying these mappings to the output specifications of P_5 and Q_5 yields, in both cases, the abstract specification

$$\textbf{assert } \alpha(\varsigma, \Delta[\kappa : \lambda]).$$

The next step in abstracting P_5 and Q_5 is to apply one of the mappings to the corresponding program. Applying the left set of transformations to P_5 yields

S_5: **begin comment** *tentative extremum schema*
  B_5: **suggest** $\lambda \in \mathbf{N}$
  $(\varsigma,i) := (\Delta[\kappa],\kappa)$
  loop L_5: **suggest** $\alpha(\varsigma,\Delta[\kappa{:}i])$, $i \in \mathbf{N}$
         until $i = \lambda$
         $i := i + 1$
         if $\Delta[i] < \varsigma$ **then** $\varsigma := \Delta[i]$ **fi**
         repeat
  E_5: **suggest** $\alpha(\varsigma,\Delta[\kappa{:}\lambda])$
  end

Had all the transformations been of variables, as are $n \Rightarrow \lambda$, $A \Rightarrow \Delta$, and $z \Rightarrow \varsigma$, then applying the transformations to all occurrences of those variables in the annotated program would of necessity have resulted in a correct schema. But since the abstraction mapping involves the transformation of constant symbols as well, viz. 0 and $\leq$, the above schema is not necessarily correct. We have, therefore, replaced the assertions with "suggestions" containing those relations that might hold for the schema as they did for the concrete program they were derived from. As we shall see, suggestions that do not turn out to be invariants can be used to further refine the schema.

The next step, therefore, is to try to determine under which conditions the suggestions are invariants. Consider the loop-exit path

> **assert** $\alpha(\varsigma,\Delta[\kappa{:}i])$, $i \in \mathbf{N}$
> **assert** $i = \lambda$
> **suggest** $\alpha(\varsigma,\Delta[\kappa{:}\lambda])$.

The first assertion is the transformed loop invariant which we shall for the moment assume was true when control was at the head of the loop; the second statement asserts that the exit test was true at that time and therefore the loop was exited; the final suggestion contains the desired output relations with respect to which we are trying to prove this schema correct. For this path to be correct, the loop invariants in the first assertion, together with the exit test of the second, must imply that the desired output relation in the suggestion holds, i.e. the verification condition for the exit path is

$$\alpha(\varsigma,\Delta[\kappa{:}i]) \ \wedge \ i\in\mathbf{N} \ \wedge \ i{=}\lambda \ \supset \ \alpha(\varsigma,\Delta[\kappa{:}\lambda]).$$

Indeed, if $\alpha(\varsigma,\Delta[\kappa{:}i])$ holds and $i{=}\lambda$, then $\alpha(\varsigma,\Delta[\kappa{:}\lambda])$ holds as well. Thus, if we can establish that $\alpha(\varsigma,\Delta[\kappa{:}i])$ and $i\in\mathbf{N}$ *are* loop invariants, the schema will be proved correct.

Next, consider the initialization path

> **assert** $\lambda\in\mathbf{N}$
> $(\varsigma,i) := (\Delta[\kappa],\kappa)$
> **suggest** $\alpha(\varsigma,\Delta[\kappa{:}i])$, $i\in\mathbf{N}$;

its verification condition may be written as

> **assert** $\lambda\in\mathbf{N}$
> **suggest** $\alpha(\Delta[\kappa],\Delta[\kappa{:}\kappa])$, $\kappa\in\mathbf{N}$.

This condition is obtained from the path by "pushing" the suggestions $\alpha(\varsigma,\Delta[\kappa{:}i])$ and $i\in\mathbf{N}$ back over the assignment $(\varsigma,i) := (\Delta[\kappa],\kappa)$, substituting the new values $\Delta[\kappa]$ and κ assigned to the variables ς and i, respectively, for occurrences of those variables in the suggestion. The subterm $\Delta[\kappa{:}\kappa]$ simplifies to $\Delta[\kappa]$, but there is no way of showing that the assertion $\lambda\in\mathbf{N}$ implies either $\alpha(\Delta[\kappa],\Delta[\kappa])$ or $\kappa\in\mathbf{N}$ (nor are there any additional abstractions that might help make the latter hold). They are, therefore, both left as preconditions for the schema to be applicable. For cosmetic purposes, we shall replace the expression $\Delta[\kappa]$ in $\alpha(\Delta[\kappa],\Delta[\kappa])$ by (the universally quantified) u and require that

$$\kappa\in\mathbf{N}$$

and

$$\alpha(u,u)$$

for all u, i.e. κ must be a nonnegative integer and α must be reflexive. This change has the effect of *strengthening* the precondition somewhat: α needs to be reflexive for "all" values u, not just $\Delta[\kappa]$. This may mean that an otherwise valid instantiation will be rejected. Note, however, that the stronger condition does hold for the two concrete instances the schema was derived from.[5]

---

[5]Had the programmer supplied the information that his program was based on the reflexivity of $\leq$, then we would be more justified in taking the reflexivity of α as a

Consider, now, the verification condition

$$\lambda \in \mathbf{N} \supset (\exists u \in \mathbf{N}) \kappa + u = \lambda$$

for termination, i.e. the exit test $i = \lambda$ will eventually hold for some value $\kappa + u$ of i, as i is being incremented from its initial value κ. Since $\lambda \in \mathbf{N}$ appears in the suggested input specification and $\kappa \in \mathbf{N}$ is already a precondition, this termination condition is equivalent to just

$$\kappa \le \lambda.$$

This too is left as a precondition.

The verification condition for the loop-body path divides into two cases, one for each possible outcome of the conditional test. In the case when the test fails, the **then**-branch is skipped, and the path taken is

> **assert** $\alpha(\varsigma, \Delta[\kappa : i])$, $i \in \mathbf{N}$
> **assert** $\sim(i = \lambda)$
> $i := i + 1$
> **assert** $\sim(\Delta[i] < \varsigma)$
> **suggest** $\alpha(\varsigma, \Delta[\kappa : i])$, $i \in \mathbf{N}$.

To verify this path we must show that the loop invariants continue to hold if the exit test is false, i is incremented, and the conditional test is false. The corresponding verification condition is

> **assert** $\alpha(\varsigma, \Delta[\kappa : i])$, $i \in \mathbf{N}$
> **assert** $\sim(i = \lambda)$
> **assert** $\sim(\Delta[i + 1] < \varsigma)$
> **suggest** $\alpha(\varsigma, \Delta[\kappa : i + 1])$, $i + 1 \in \mathbf{N}$.

Clearly, if $i \in \mathbf{N}$, then $i + 1 \in \mathbf{N}$, as well. We must also show $\alpha(\varsigma, \Delta[\kappa : i + 1])$. But it is already assumed that $\alpha(\varsigma, \Delta[\kappa : i])$; all that remains to show is $\alpha(\varsigma, \Delta[i + 1])$. The only relevant assumption relating ς and $\Delta[i + 1]$ is $\sim(\Delta[i + 1] < \varsigma)$. Of course, since α is an abstract predicate, there is no reason for $\sim(\Delta[i + 1] < \varsigma)$ to imply $\alpha(\varsigma, \Delta[i + 1])$. Accordingly, we look for an extension to the abstraction mapping that

---

precondition. Without that information, one might still make simplifying assumptions, based on appropriate heuristic notions of "simplicity," as long as the simplified conditions hold for all the given programs.

will make this condition hold:

$$\sim(\Delta[i+1] < \varsigma) \quad \Rightarrow \quad \alpha(\varsigma, \Delta[i+1]).$$

Negating the two sides (i.e. inverting) gives

$$\Delta[i+1] < \varsigma \quad \Rightarrow \quad \sim\alpha(\varsigma, \Delta[i+1]),$$

for which it is only necessary (by imitation) to transform[6]

$$u < v \quad \Rightarrow \quad \sim\alpha(v, u).$$

Applying this additional transformation to (the loop body of) the schema, we get the conditional statement

if $\sim\alpha(\varsigma, \Delta[i])$ **then** $\varsigma := \Delta[i]$ **fi**.

Now the verification condition for the case when the test is false carries through, and it remains only to verify the case when $\sim\alpha(\varsigma, \Delta[i])$ is true:

assert $\alpha(\varsigma, \Delta[\kappa:i])$, $i \in \mathbf{N}$
assert $\sim(i = \lambda)$
$i := i + 1$
assert $\sim\alpha(\varsigma, \Delta[i])$
$\varsigma := \Delta[i]$
suggest $\alpha(\varsigma, \Delta[\kappa:i])$, $i \in \mathbf{N}$.

The verification condition for this case is

assert $\alpha(\varsigma, \Delta[\kappa:i])$, $i \in \mathbf{N}$
assert $\sim(i = \lambda)$
assert $\sim\alpha(\varsigma, \Delta[i+1])$
suggest $\alpha(\Delta[i+1], \Delta[\kappa:i+1])$, $i+1 \in \mathbf{N}$.

Since we are already assuming that α is reflexive, in which case $\alpha(\Delta[i+1], \Delta[i+1])$, we need only ascertain $\alpha(\Delta[i+1], \Delta[\kappa:i])$. Since there is no way to prove this to hold for all α (nor is there any way to extend the analogy so that it will), it is also left as a precondition:

$$\alpha(\varsigma, \Delta[\kappa:i]) \ \wedge \ i \in \mathbf{N} \ \wedge \ \sim(i = \lambda) \ \wedge \ \sim\alpha(\varsigma, \Delta[i+1])$$
$$\supset \ \alpha(\Delta[i+1], \Delta[\kappa:i]).$$

Again, for cosmetic purposes we will require instead (the more restrictive

---

[6]Note the importance of not having simplified $\sim(\Delta[i+1] < \varsigma)$ to $\Delta[i+1] \geq \varsigma$ in the verification condition, so as not to lose sight of the symbols as they appear in the code.

condition)

$$\alpha(w,u) \; \wedge \; \sim\alpha(w,v) \; \supset \; \alpha(v,u)$$

for all u, v, and w.[7]

Applying the complete abstraction mapping

$$
\begin{array}{rcl}
0 & \Rightarrow & \kappa \\
n & \Rightarrow & \lambda \\
u \leq v & \Rightarrow & \alpha(u,v) \\
A & \Rightarrow & \Delta \\
z & \Rightarrow & \varsigma \\
u < v & \Rightarrow & \sim\alpha(v,u)
\end{array}
$$

to program P_5, our final version of the schema is

```
S₅: begin comment extremum schema
  B₅: assert α(u,u), α(w,u)∧~α(w,v)⊃α(v,u),
          κ∈N, λ∈N, κ≤λ
  (ς,i) := (Δ[κ],κ)
  loop L₅: assert α(ς,Δ[κ:i]), i∈N
      until i=λ
      i := i+1
      if ~α(ς,Δ[i]) then ς := Δ[i] fi
      repeat
  E₅: assert α(ς,Δ[κ:λ])
  end
```

Now that we have verified the conditions for each of the paths in the
program, the suggestions have been replaced by assertions. The preconditions are given in the input assertion; any instantiation that satisfies
them is guaranteed to yield a correct program. Of course, the predicate
α that appears in the schema should be instantiated to a primitive predicate of the target language, or else it will need to be replaced for the

---

[7]Had we not simplified the reflexivity precondition, then at this point we would obtain an additional condition $\alpha(\Delta[i+1],\Delta[i+1])$.

program to be executable. Likewise, the constants Δ, κ, and λ must be replaced with primitives, or code prepended to set them.[8]

Recall that the given output specification of program P_5 was

assert $z \leq A[0{:}n]$.

Actually, a more realistic requirement specification for a search for a minimum would have included the requirement $z \in A[0{:}n]$, i.e. not only should z be no larger than any element of $A[0{:}n]$, but it should also be one of those elements. If we apply our abstraction mapping to this additional specification, we get the abstract relation $\varsigma \in \Delta[\kappa{:}\lambda]$, which indeed holds true for our schema. In other ways too, the form in which the specifications of programs are presented can influence the ease with which they can be abstracted. Had P_5 and Q_5, for example, been specified to achieve $z=min\{A[0{:}n]\}$ and $B[p]=max\{B[b{:}c]\}$, respectively, then the analogy between $\leq$ and $\geq$ would have remained to be discovered from the analysis of the verification conditions. If the two programs were each specified in a different manner, one would first have to find equivalent, similar specifications, before proceeding.[9]

Programs for finding the position or value of the minimum or maximum of an array (or of other functions with integer domain) are valid instantiations of this schema. For instance, say we want to find the position m of the minimum of some function f for the first fifty odd integers. Comparing this goal,

achieve $(\forall u \in [1{:}100]) \ (odd(u) \supset f(m) \leq f(u))$ **varying** m

(meaning that m should be such that $f(m)$ is not greater than $f(u)$ for every integer u in the range $[1{:}100]$ that is odd),[10] with the abstract

---

[8]Had we applied the corresponding abstraction mapping to Q_5 instead of to P_5, a superficially different, but equally valid, schema would have resulted.

[9]Had the specifications been $(\forall u \in \mathbf{Z}) \ (0 \leq u \leq n \supset z \leq A[u])$ and $(\forall u \in \mathbf{Z})$ $(a \leq u \leq b \supset B[p] \geq B[u])$, then the analogy between $\leq$ and $\geq$ would have to be restrained to apply only to *some* occurrences of $\leq$. To determine which occurrences in the program ought to be transformed requires the kind of more detailed analysis that we attempt in the next example.

[10]One could just as well apply the schema to the alternative specification

output specification of the schema

$$\textbf{assert } \alpha(\varsigma,\Delta[\kappa\!:\!\lambda]),$$

that is

$$\textbf{assert } (\forall u \in [\kappa\!:\!\lambda])\ \alpha(\varsigma,\Delta[u]),$$

suggests the (imitating) mappings

$$
\begin{array}{rcl}
\kappa & \Rightarrow & 1 \\
\lambda & \Rightarrow & 100 \\
\alpha(\varsigma,\Delta[u]) & \Rightarrow & odd(u) \supset f(m) \leq f(u).
\end{array}
$$

The last of these mappings can be accomplished by the (projecting and imitating) mappings

$$
\begin{array}{rcl}
\Delta[u] & \Rightarrow & u \\
\alpha(v,w) & \Rightarrow & odd(w) \supset f(v) \leq f(w) \\
\varsigma & \Rightarrow & m.
\end{array}
$$

Applying this instantiation mapping to the five preconditions

$$
\begin{array}{c}
\kappa \in \mathbf{N} \\
\lambda \in \mathbf{N} \\
\kappa \leq \lambda \\
\alpha(u,u) \\
\alpha(w,u) \ \wedge \ \sim\!\alpha(w,v) \ \supset \ \alpha(v,u)
\end{array}
$$

yields

$$
\begin{array}{c}
1 \in \mathbf{N} \\
100 \in \mathbf{N} \\
1 \leq 100 \\
odd(u) \ \supset \ f(u) \leq f(u) \\
[odd(u) \supset f(w) \leq f(u)] \ \wedge \ \sim\![odd(v) \supset f(w) \leq f(v)] \\
\supset \ [odd(u) \supset f(v) \leq f(u)].
\end{array}
$$

The first three conditions are obviously true; the fourth holds since $\leq$ is reflexive; the last follows from transitivity. Applying the instantiation mapping to the schema yields the program

---

$(\forall u \in [1\!:\!50])\ f(m) \leq f(2 \cdot u - 1).$

R_5: **begin comment** *function minimum program*
  $(m,i) := (1,1)$
  loop L_5: **assert** $(\forall u \in [1{:}i])\ (odd(u) \supset f(m) \leq f(u)),\ i \in \mathbf{N}$
    until $i = 100$
    $i := i + 1$
    if $odd(i) \wedge f(m) > f(i)$ **then** $m := i$ **fi**
    repeat
  E_5: **assert** $(\forall u \in [1{:}100])\ (odd(u) \supset f(m) \leq f(u))$
end

where the transformed conditional test $\sim(odd(i) \supset f(m) \leq f(i))$ simplified to $odd(i) \wedge f(m) > f(i)$. Since the preconditions were satisfied by the instantiation, this program is guaranteed to be correct. Further improvements to the instantiated program could be made (e.g. incrementing i by two with each iteration) by a straightforward series of correctness-preserving program transformations.

4.3.2. Binary Search

For this example, we suppose that the programmer has supplied detailed comments on the reasoning he employed in constructing his programs. Such additional information may help in extending the analogy between the programs and arriving at a correct schema. It may also allow the localization of transformations to relevant occurrences of symbols and will help avoid unnecessarily strict preconditions.[11]

---

[11]Such information is not essential for this example to succeed; it would be enough were loop invariants supplied as in the previous example.

The following two programs both use the binary-search technique:

```
assert 0 ≤ c < d, e > 0, s/2 ∈ 1/2^N, q ≥ 0 in
P₁: begin comment real-division program
    B₁: assert 0 ≤ c < d, e > 0
    purpose |c/d−q| < e
        purpose q ≤ c/d, c/d < q + s, s ≤ e
        (q,s) := (0,2)
        loop L₁: assert q ≤ c/d, c/d < q + s
            until s ≤ e
            purpose q ≤ c/d, c/d < q + s, 0 < s < s_{L₁}
            s := s/2
            if d·(q + s) ≤ c then q := q + s fi
        repeat
    assert q ≤ c/d, c/d < q + s, s ≤ e
    E₁: assert |c/d−q| < e
    end
```

```
assert a > 1, t > 0, r ≥ 0 in
Q₁: begin comment real square-root program
    B₁: assert a > 1, t > 0
    purpose |√a −r| < t
        purpose r ≤ √a, √a < w, w ≤ r + t
        (r,w) := (1,a)
        loop L₁: assert r ≤ √a, √a < w
            while w − r > t
            purpose r ≤ √a, √a < w, 0 < w−r < w_{L₁}−r_{L₁}
            p := (w + r)/2
            if p² ≤ a   then   r := p
                        else   w := p   fi
        repeat
    assert r ≤ √a, √a < w, w ≤ r + t
    E₁: assert |√a −r| < t
    end
```

The first uses that technique to find the quotient q of two nonnegative real numbers c and d, $c < d$, within a given positive tolerance e.[12] The

---

[12] That program—taken from Section 2.4—initializes s to 2 (instead of to the better choice 1) to illustrate a point later on.

second finds the square-root r of the real number a, $a > 1$, within positive tolerance t. Both programs are annotated with comments in the form of **purpose** statements, listing relations that the programmer intended for the code that follows to achieve.

Comparing the output specification

$$E_1: \textbf{assert } |c/d-q| < e$$

of P_1 with the specification

$$E_1: \textbf{assert } |\sqrt{a} - r| < t$$

of Q_1, suggests the analogy

$$
\begin{array}{ccc}
q & \Longleftrightarrow & r \\
u/d & \Longleftrightarrow & \sqrt{u} \\
c & \Longleftrightarrow & a \\
e & \Longleftrightarrow & t.
\end{array}
$$

Now we will work our way from the specification, through the programmer's comments, to extend and refine this analogy. To begin with, the programmer achieved the output specification $|c/d-q| < e$ of P_1 by decomposing it into the three conjunctive subgoals given in his comment

$$\textbf{purpose } q \leq c/d, \ c/d < q + s, \ s \leq e.$$

(The last conjunct became the exit test of the loop and the other two became loop invariants.) Abstracting these subgoals, by applying the abstraction mapping

$$
\begin{array}{ccc}
q & \Rightarrow & \xi \\
u/d & \Rightarrow & \rho(u) \\
c & \Rightarrow & \chi \\
e & \Rightarrow & \epsilon
\end{array}
$$

for P_1, gives the abstract subgoals

$$\textbf{purpose } \xi \leq \rho(\chi), \ \rho(\chi) < \xi + s, \ s \leq \epsilon,$$

which indeed imply (and are all needed to imply) the desired output specification of the schema. Similarly, the goal $|\sqrt{a} - r| < t$ of Q_1 was reduced to the three subgoals

purpose $r \leq \sqrt{a}$, $\sqrt{a} < w$, $w \leq r + t$.

Applying the corresponding abstraction mapping

$$\begin{array}{ccc} r & \Rightarrow & \xi \\ \sqrt{u} & \Rightarrow & \rho(u) \\ a & \Rightarrow & \chi \\ t & \Rightarrow & \epsilon \end{array}$$

for Q_1 to these subgoals yields

purpose $\xi \leq \rho(\chi)$, $\rho(\chi) < w$, $w \leq \xi + \epsilon$.

These are not however identical with the subgoals for P_1; to make them equivalent requires extending the analogy with

$$\xi + s \quad \Longleftrightarrow \quad w.$$

If we let η be their abstract counterpart, then

$$s \quad \Rightarrow \quad \eta - \xi$$

must be added to the abstraction mapping of P_1 and

$$w \quad \Rightarrow \quad \eta$$

to the mapping for Q_1.

The initialization of the division program was correct since $a < b$ and, therefore, when q is set to 0 and s to 2, we can be sure that $q \leq c/d < q + s$; the correctness of the initialization of the square-root program follows from the fact that $a > 1$, $r = 1$, and $w = a$ imply $r \leq \sqrt{a} < w$. This suggests extending the analogy between the two programs with

$$\begin{array}{ccc} 0 & \Longleftrightarrow & 1 \\ 2 & \Longleftrightarrow & a \end{array}$$

and extending the abstraction mappings with

$$\begin{array}{ccccccl} 0 & \Rightarrow & o & \Leftarrow & 1 & \textit{(initialization)} \\ 2 & \Rightarrow & \iota & \Leftarrow & a & \textit{(initialization)}. \end{array}$$

These transformations are only applied to the respective initialization parts of the programs. Other occurrences of 2 in P_1 should *not* be transformed, as they do not affect the correctness of the initialization. The resultant abstract condition for the correctness of the initialization is

$$o \leq \rho(\chi) < \iota.$$

This is our first precondition.

The purpose that the programmer had in mind for the loop body of P_1 was

purpose $q \leq c/d,\ c/d < q+s,\ 0 < s < s'$,

where s' denotes the value of the variable s when control was last at the head of the loop. In other words, the loop body reachieves the invariants while making progress towards the exit test by decreasing s (some minimal amount—to ensure termination). Abstracting this goal, or the corresponding one for Q_1, yields

purpose $\xi \leq \rho(\chi),\ \rho(\chi) < \eta,\ 0 < \eta - \xi < \eta' - \xi'$

To achieve the last conjunct of this loop-body goal, the division program decreases s. Then, to achieve the remaining two conjuncts, a conditional statement with the

purpose $q \leq c/d,\ c/d < q+s$

(for the decreased value of s) is introduced. Since only the value of s has been changed so far by the loop body, the relation $q \leq c/d$ still holds, but $c/d < q+s$ might not. The program checks if the latter still holds by testing the equivalent condition $d \cdot (q+s) \leq c$. It is here, however, that the consistency of the schema breaks down: the transformed test $d \cdot \eta \leq \chi$ will not determine whether the abstract relation $\rho(\chi) < \eta$ holds. Similarly, finding $p^2 \leq \chi$ to be false and assigning $\eta := p$, obtained by abstracting Q_1, does not achieve $\rho(\chi) < \eta$.

There is, nevertheless, an analogy between the two transformed conditionals. Where the one achieves $d \cdot \eta > \chi$, the other achieves $\eta^2 > \chi$. Accordingly, the analogy between P_1 and Q_1 can be extended with

$$d \cdot u \quad \Rightarrow \quad \sigma(u) \quad \Leftarrow \quad u^2 \quad \textit{(conditional test)}$$

Note that these transformations should only be applied to the conditional test, since that was where the problem arose. This is another example of a *localized* transformation.

Now, for the transformed conditional statement to have the desired effect, we need $\sim(\sigma((\eta+\xi)/2)\leq\chi)$ to imply $\rho(\chi)<(\eta+\xi)/2$, or more generally

$$\sigma(u)>v \supset u>\rho(v).$$

This becomes a precondition. We must also determine the conditions under which the **then**-branch of the conditional is correct. That case was correct in the division program because $d\cdot(q+s/2)\leq c$ implies $q+s/2\leq c/d$, while $c/d<q+s$ implies $c/d<q+s/2+s/2$. For the abstracted schema, then, we need $\sigma((\eta+\xi)/2)\leq\chi$ to imply $(\eta+\xi)/2\leq\rho(\chi)$ and $\rho(\chi)<\eta$ to imply $\rho(\chi)<2\cdot(\eta+\xi)/2-\xi$. The second implication obviously holds; the first yields the additional precondition

$$\sigma(u)\leq v \supset u\leq\rho(v).$$

Combined with the previous precondition, we have

$$\sigma(u)\leq v \equiv u\leq\rho(v),$$

which holds, in particular, if σ is the inverse of a monotonic function ρ, i.e. if $\rho(\sigma(u))=u$ and $u\leq v\equiv\rho(u)\leq\rho(v)$ for all u and v.

Putting everything together, the complete abstraction mapping for P_1 is

$$
\begin{array}{rcl}
q & \Rightarrow & \xi \\
u/d & \Rightarrow & \rho(u) \\
c & \Rightarrow & \chi \\
e & \Rightarrow & \epsilon \\
s & \Rightarrow & \eta-\xi \\
0 & \Rightarrow & o \qquad \textit{(initialization)} \\
2 & \Rightarrow & \iota \qquad \textit{(initialization)} \\
d\cdot u & \Rightarrow & \sigma(u) \qquad \textit{(conditional test),}
\end{array}
$$

the last three of which are localized as indicated.

There are, however, problems with applying the transformation $s\Rightarrow\eta-\xi$ to the statements of P_1. If we apply it in a straightforward manner to the initialization assignment

$$(\xi,s) := (0,2),$$

then we would obtain $(\xi,\eta-\xi) := (0,2)$. But the assignment $\eta-\xi := 2$

is illegal, as an expression may not appear on the left-hand side of an assignment. Nevertheless, the desired effect of making the difference between the new values of η and ξ equal to 2, i.e. for $\eta-0=2$ (0 is the new value of ξ), can be achieved by the legal assignment $\eta := 2+0$. We have, then,

$$(\xi,\eta) := (0,2).$$

Similarly, the loop-body assignment

$$s := s/2$$

transforms to $\eta-\xi := (\eta-\xi)/2$. To make it legal, the ξ must be transposed to the right-hand side; the resultant assignment is

$$\eta := (\eta+\xi)/2.$$

We are not yet finished, however, as the value of the difference $\eta-\xi$ also changes whenever ξ is assigned to. Accordingly, we must look at the **then**-branch assignment

$$\xi := \xi+s,$$

viewing it as $(\xi,s) := (\xi+s,s)$, where we have explicitly included a dummy assignment to the variable s. Transforming this augmented assignment gives $(\xi,\eta-\xi) := (\xi+\eta-\xi,\eta-\xi)$. Thus, ξ should get the value η, while η should get the value $\eta-\xi$ plus the *new* value of ξ, which happens to be the *old* value of η. The appropriate legal assignment is, therefore,

$$(\xi,\eta) := (\eta,2{\cdot}\eta-\xi),$$

and the abstracted program is

$$
\begin{aligned}
&(\xi,\eta) := (o,\iota) \\
&\textbf{loop } L_1\text{: } \textbf{assert } \xi \le \rho(\chi),\ \rho(\chi) < \eta \\
&\quad \textbf{until } \eta-\xi \le \epsilon \\
&\quad \eta := (\eta+\xi)/2 \\
&\quad \textbf{if } \sigma(\eta) \le \chi \textbf{ then } (\xi,\eta) := (\eta,2{\cdot}\eta-\xi) \textbf{ fi} \\
&\quad \textbf{repeat.}
\end{aligned}
$$

This schema may be improved slightly by applying the correctness-preserving global transformation

$$\eta \ \Rightarrow \ \xi + \eta.$$

This is like having originally abstracted s into η, rather than into $\eta - \xi$, and, consequently, the schema

S_1: **begin comment** *binary-search schema*
  B_1: **assert** $\sigma(u) \leq v \equiv u \leq \rho(v)$, $o \leq \rho(\chi) < \iota$, $\epsilon > 0$
  $(\xi, \eta) := (o, \iota - o)$
  loop L_1: **assert** $\xi \leq \rho(\chi)$, $\rho(\chi) < \xi + \eta$
      until $\eta \leq \epsilon$
      $\eta := \eta/2$
      if $\sigma(\xi + \eta) \leq \chi$ **then** $\xi := \xi + \eta$ **fi**
      repeat
  E_1: **assert** $|\rho(\chi) - \xi| < \epsilon$
  end

that results is more similar to P_1. It is a general-purpose program schema that performs a binary-search for the value ξ of an invertible monotonic function ρ, at the point χ, within a tolerance ϵ. In Section 2.7, we saw how this schema may be applied to the computation of the square-root of an integer.

4.3.3. Associative Recursion

Abstraction may also be used for the design of correctness-preserving program transformations. For example, one can first write—and verify—iterative versions of several recursive programs; by abstracting those programs—and their proofs—a more general recursion-to-iteration transformation schema can be obtained. As we shall see, very little information is gleaned just by a comparison of output specifications; instead, most of the analogy needs to be derived from the recursive and iterative programs.

Recursive and iterative versions of factorial and summation programs were given in Section 4.2.4, where we described how a recursion-elimination transformation can be derived from them. In this section, we consider the same recursive programs, but different iterative ones, and illustrate the derivation of an alternative program transformation. The two iterative programs are

P_6: **begin comment** *iterative factorial program*
  B_6: **assert** $n \in \mathbf{N}$
  $(m,y) := (1,n)$
  loop L_6: **assert** $m \cdot y! = n!$, $y \in \mathbf{N}$
      until $y = 0$
      $(m,y) := (m \cdot y, y - 1)$
      repeat
  E_6: **assert** $m = n!$
  end

and

Q_6: **begin comment** *iterative summation program*
  B_6: **assert** $k \in \mathbf{Z}$, $l \in \mathbf{Z}$, $k \leq l$
  $(s,j) := (0,k)$
  loop L_6: **assert** $s = \sum_{i=k}^{j-1} f(i)$, $j \in \mathbf{Z}$
      until $j = l + 1$
      $(s,j) := (s + f(j), j + 1)$
      repeat
  E_6: **assert** $s = \sum_{i=k}^{l} f(i)$
  end

Our object is to derive a schema for computing recursive functions that are like factorial and summation. Matching the two output specifications

$$\textbf{assert } m = n!$$

and

$$\textbf{assert } s = \sum_{i=k}^{l} f(i)$$

suggests as one possible analogy

$$
\begin{array}{ccc}
m & \Longleftrightarrow & s \\
n & \Longleftrightarrow & k \\
u! & \Longleftrightarrow & \sum_{i=u}^{l} f(i).
\end{array}
$$

The two output variables m and s generalize to the abstract output variable μ; the input variables n and k generalize to ν; the two functions $u!$ and $\sum_{i=u}^{l} f(i)$ generalize to an abstract function variable $\phi(u)$. The abstracted output specification is accordingly

$$\textbf{suggest } \mu = \phi(\nu).$$

To complete the analogy, we must examine the programs in more detail. First, we should compare the recursive versions of these two programs and extend the analogy; then, we should concentrate on the iterative versions and see what conditions must be placed on the underlying recursive functions for the program transformation to be valid. A comparison of the recursive programs,

$P_6'(m,n)$: **begin comment** *recursive factorial program*
   if $n = 0$ **then** $m := 1$
             else $P_6'(m, n-1)$
                     $m := m \cdot n$ **fi**
  E_6': **assert** $m = n!$
  end

for factorial, and

$Q_6'(k,s)$: **begin comment** *recursive summation program*
   if $k = l+1$ **then** $s := 0$
                else $Q_6'(k+1, s)$
                     $s := s + f(k)$ **fi**
  E_6': **assert** $s = \sum_{i=k}^{l} f(i)$
  end

for summation, suggests the additional analogies

$$
\begin{array}{ccccc}
0 & \Rightarrow & \theta & \Leftarrow & l+1 \\
1 & \Rightarrow & \omega & \Leftarrow & 0 \\
u-1 & \Rightarrow & \delta(u) & \Leftarrow & u+1 \\
u \cdot v & \Rightarrow & \psi(u,v) & \Leftarrow & u+v \\
v & \Rightarrow & \sigma(v) & \Leftarrow & f(v).
\end{array}
$$

Note the use of a projecting mapping between $f(u)$ in Q_6' and u in P_6', on account of which it would be dangerous to apply the abstraction mapping to P_6'.[13] Applying the mapping

---

[13]The alternative analogy pursued in Section 4.2.4 would lead, in this example, to unduly complex preconditions, which would hamper efforts to instantiate the schema.

$$
\begin{array}{rcl}
s & \Rightarrow & \mu \\
k & \Rightarrow & \nu \\
\sum_{i=u}^{l} f(i) & \Rightarrow & \phi(u) \\
l+1 & \Rightarrow & \theta \\
0 & \Rightarrow & \omega \\
v+1 & \Rightarrow & \delta(u) \\
u+v & \Rightarrow & \psi(u,v) \\
f(v) & \Rightarrow & \sigma(v)
\end{array}
$$

to the recursive version Q_6' of summation, yields the recursive schema

---

$R_7(\nu,\mu)$: **begin comment** *recursion schema*
   if $\nu=\theta$ **then** $\mu := \omega$
         else $R_7(\delta(\nu),\mu)$
              $\mu := \psi(\mu,\sigma(\nu))$ **fi**
  E_7: **assert** $\mu=\phi(\nu)$
end

Applying the same mapping to the iterative version Q_6 yields the schema

---

S_7': **begin comment** *abstracted summation program*
  B_7': **suggest** $\nu\in\mathbf{Z}$, $l\in\mathbf{Z}$, $\nu\leq l$
  $(\mu,j) := (\omega,\nu)$
  loop L_7': **suggest** $\mu=\sum_{i=\nu}^{j-1}\sigma(i)$, $j\in\mathbf{Z}$
      until $j=\theta$
      $(\mu,j) := (\psi(\mu,\sigma(j)),\delta(j))$
      repeat
  E_7': **suggest** $\mu=\phi(\nu)$
end

The problem at this point shows up in the loop invariant

$$
L_7': \textbf{ suggest } \mu=\sum_{i=\nu}^{j-1}\sigma(i), \; j\in\mathbf{Z}
$$

of the above iterative schema. With it, there is no way that the verification conditions will hold. This problem stems from the fact that the form of the given invariant for the summation program relies on properties of sums. A clue to resolving this dilemma is to try and use instead the abstracted loop invariant

$$L_7: \quad \textbf{suggest} \quad \psi(\mu,\phi(y))=\phi(\nu), \; y \in \mathbf{N}$$

of the iterative factorial program, obtained by applying the mapping

$$
\begin{array}{rcl}
m & \Rightarrow & \mu \\
n & \Rightarrow & \nu \\
u! & \Rightarrow & \phi(u) \\
0 & \Rightarrow & \theta \\
1 & \Rightarrow & \omega \\
u-1 & \Rightarrow & \delta(u) \\
u \cdot v & \Rightarrow & \psi(u,v)
\end{array}
$$

to

$$L_6: \quad \textbf{assert} \quad m \cdot y! = n!, \; y \in \mathbf{N}$$

(avoiding the certainly overzealous abstraction $v \Rightarrow \sigma(v)$). Comparing $y \in \mathbf{N}$ with $j \in \mathbf{Z}$ suggests expanding the analogy with

$$
\begin{array}{ccccc}
y & \Rightarrow & \xi & \Leftarrow & j \\
\mathbf{N} & \Rightarrow & \Xi & \Leftarrow & \mathbf{Z}.
\end{array}
$$

That gives us the suggested invariant

$$L_7: \quad \textbf{suggest} \quad \psi(\mu,\phi(\xi))=\phi(\nu), \; \xi \in \Xi$$

for S_7. With this as the loop invariant, we proceed to examine the schema's verification conditions. The initialization condition of S_7 is

$$\textbf{suggest} \quad \psi(\omega,\phi(\nu))=\phi(\nu), \; \nu \in \Xi.$$

Making

$$\nu \in \Xi$$

into a precondition, as well as the (simplified) condition

$$\psi(\omega,u) = u,$$

will insure the correctness of the initialization path.

The loop-exit verification condition is

> **assert** $\psi(\mu,\phi(\xi))=\phi(\nu)$, $\xi\in\Xi$
> **assert** $\xi=\theta$
> **suggest** $\mu=\phi(\nu)$.

For simplicity, we replace this condition with

$$\psi(u,\phi(\theta))=u.$$

For the loop-body path, we have the condition

> **assert** $\psi(\mu,\phi(\xi))=\phi(\nu)$, $\xi\in\Xi$
> **assert** $\sim(\xi=\theta)$
> **suggest** $\psi(\psi(\mu,\sigma(\xi)),\phi(\delta(\xi)))=\phi(\nu)$, $\delta(\xi)\in\Xi$,

for which we shall use the two simpler preconditions

$$v\in\Xi \wedge v\neq\theta \supset \delta(v)\in\Xi$$

and

$$v\neq\theta \supset \psi(u,\phi(v))=\psi(\psi(u,\sigma(v)),\phi(\delta(v))).$$

Finally, the verification condition for the termination of the abstracted program is

$$(\exists u\in\mathbf{N})\ \delta^u(\nu)=\theta.$$

This becomes our last precondition.

The complete abstraction mappings are

m	$\Rightarrow$	μ	$\Leftarrow$	s
n	$\Rightarrow$	ν	$\Leftarrow$	k
$u!$	$\Rightarrow$	$\phi(u)$	$\Leftarrow$	$\sum_{i=u}^{l} f(i)$
0	$\Rightarrow$	θ	$\Leftarrow$	$l+1$
1	$\Rightarrow$	ω	$\Leftarrow$	0
$u-1$	$\Rightarrow$	$\delta(u)$	$\Leftarrow$	$u+1$
$u\cdot v$	$\Rightarrow$	$\psi(u,v)$	$\Leftarrow$	$u+v$
v	$\Rightarrow$	$\sigma(v)$	$\Leftarrow$	$f(v)$
y	$\Rightarrow$	ξ	$\Leftarrow$	j
$\mathbf{N}$	$\Rightarrow$	Ξ	$\Leftarrow$	$\mathbf{Z}$

Applying the abstraction transformations to Q_6, and collecting all the preconditions, we obtain the schema

S_7: **begin comment** *associative recursion schema*
  B_7: **assert** $\psi(\omega,u)=u$, $\psi(u,\phi(\theta))=u$,
            $\nu \in \Xi$, $v \in \Xi \wedge v \neq \theta \supset \delta(v) \in \Xi$,
            $v \neq \theta \supset \psi(u,\phi(v))=\psi(\psi(u,\sigma(v)),\phi(\delta(v)))$,
            $(\exists u \in N)\ \delta^u(\nu)=\theta$
  $(\mu,\xi) := (\omega,\nu)$
  loop L_7: **assert** $\psi(\mu,\phi(\xi))=\phi(\nu)$, $\xi \in \Xi$
      until $\xi=\theta$
      $(\mu,\xi) := (\psi(\mu,\sigma(\xi)),\delta(\xi))$
      repeat
  E_7: **assert** $\mu=\phi(\nu)$
end

In this manner we have obtained a general schema for computing a function $\phi(\nu)$. It applies to recursive programs of the form R_7 that compute a function $\phi(x)$, such that $\phi(x)=\psi(x,\phi(\delta(x)))$ when x is not θ, $\phi(\theta)=\omega$, and ω is an identity element of an associative and commutative function ψ. This schema is similar to one of the recursion-to-iteration transformations of [DarlingtonBurstall76].

To see how the schematic transformation

R_7: *recursion schema*
S_7: *associative recursion schema*

may be applied to another problem, consider the recursive program

$P_5'(n,z)$: **begin comment** *recursive array-minimum program*
  B_5': **assert** $n \in N$
  if $n=0$ **then** $z := \infty$
             else $P_5'(n-1,z)$
                   $z := min(z,A[n])$ **fi**
  E_5': **assert** $z=min\{A[1:n]\}$
end

where ∞ is greater than "any" array value, so it serves as the minimum of an empty array. We wish to transform this program into an iterative one that finds the smallest element.

An initial comparison of this program's output specification with the specification $\mu = \phi(\nu)$ of our schema, suggests the possible instantiation

$$\phi(u) \;\Rightarrow\; min\{A[1:u]\}$$
$$\mu \;\Rightarrow\; z$$
$$\nu \;\Rightarrow\; n.$$

A comparison of P_5' with the recursive pattern R_7 suggests the additional instantiations

$$\theta \;\Rightarrow\; 0$$
$$\omega \;\Rightarrow\; \infty$$
$$\delta(u) \;\Rightarrow\; u-1$$
$$\psi(u,v) \;\Rightarrow\; min(u,v)$$
$$\sigma(v) \;\Rightarrow\; A[v]$$
$$\Xi \;\Rightarrow\; \mathbf{N}.$$

It remains to check the validity of the preconditions. Applying the instantiation mapping to the six of them gives

$$min(\infty,u) = u$$
$$min(u,min\{A[1:0]\}) = u$$
$$n \in \mathbf{N}$$
$$v \in \mathbf{N} \wedge v \neq 0 \supset v-1 \in \mathbf{N}$$
$$v \neq 0 \supset min(u,\{A[1:v]\}) = min(min(u,A[v]),min\{A[1:v-1]\})$$
$$(\exists u \in \mathbf{N}) \; \nu - u = 0$$

Since they are all true, the following instantiation of S_7 is a correct iterative program for array-minimum:

```
P₅: begin comment iterative array−minimum program
  B₅: assert n ∈ N
  (z,ξ) := (∞,n)
  loop L₅: assert min(z,min{A[1:ξ]})=min{A[1:n]},  ξ∈N
      until ξ=0
      (z,ξ) := (min(z,A[ξ]),ξ−1)
      repeat
  E₅: assert z=min{A[1:n]}
  end
```

4.4. Discussion

We have presented several examples demonstrating a methodology for deriving an abstract schema from a given set of concrete programs. Once derived, the schema may be applied to solve new problems by instantiating the abstract entities of the schema with concrete elements from the problem domain. We have also seen how this methodology can be used to derive correctness-preserving program transformations and to guide their application.

Abstraction and instantiation complement other methods of program transformation. When faced with the task of developing a new program (or subprogram) to meet a set of specifications, a programmer ought to first search for an applicable schema. After instantiating that schema appropriately, various other transformations may be necessary to satisfy remaining specifications or increase efficiency. When no applicable schema can be found, one might still be able to find a program solving an analogous problem, and modify it, as in the previous chapter. Those two programs together could then be used to formulate a schema for future use. Using a schema has the advantage over modifying a related program in that correctness is ensured by satisfying the preconditions. The extraction of appropriate preconditions, however, is what makes abstraction more complex a task than either modification or instantiation.

The abstraction process begins with a set of programs. How many programs? On the one hand, it is conceivable that one might abstract a single program by somehow choosing specification symbols to be abstracted and proceeding to examine the program's proof. This would be better than attempting to formulate a schema from scratch (something we do in the next chapter), since the proof provides a means of specifying necessary properties of the abstract objects. On the other hand, using more programs reduces arbitrariness by insisting that only analogous parts be abstracted, on the presumption that they play similar roles in the different programs. It is the analogy, then, that suggests which "parameters" of a program are worthy of generalization. If additional programs become available, then there would be no reason not to attempt to generalize the derived schemata even further. By the same token, two schemata could be themselves abstracted. In such a manner, hierarchies of schemata would be created, each more abstract than its predecessors, as, for example, the binary-search schema is an instance of a more general "divide and conquer" stratagem. The related question of which programs are *worthy* of abstraction is difficult to answer, let alone automate! One criterion might be the frequency with which a program is modified for tasks other than that which it was originally intended for. That is, the more a program is "cannibalized," the more worthy its skeleton of "immortalization." How many schemata are in the typical programmer's repertoire? That may be as hard to say as identifying all the archetypical motifs used by comedians.

There are some problems inherent in the use of analogies for program abstraction and instantiation. These include "hidden" analogies, "misleading" analogies, "incomplete" analogies, and "overzealous" analogies. Hidden analogies arise when given specifications (of existing programs in the case of abstraction, or of an abstract schema and concrete problem in the case of instantiation) that are to be compared with one another have little syntactically in common. It takes only small variations to make semantically similar objects appear unrelated. Since the pattern-matching ideas that we have employed are syntax-based, when the specifications are not syntactically similar, the underlying analogy would be disguised. In such a situation, before an analogy could be found, it would be necessary to rephrase the specifications in an equivalent manner, one that brings out the similarities. This is clearly a difficult problem in its own right; in general some form of "means-end" analysis ([NewellSimon72]) seems appropriate.

At the opposite extreme, a syntactic analogy may be misleading. The same symbol may appear in the specifications of two programs, yet play nonanalogous roles in them. Two programs might even have the exact same specifications, but employ totally different methods of solution. Situations such as these would be detected in the course of analyzing the correctness conditions for the abstracted programs. We have also seen how the proof of correctness of a program can be used to help avoid overzealously applying transformations to unrelated parts of a program, to complete an analogy between two programs (only part of which was found by a comparison of specifications), and to derive preconditions for applicability of a schema. For all these reasons, it is important that programs be accompanied by proofs. That would be the case if programs are developed formally and systematically from specifications, as described in the next chapter, or if invariants are extracted from existent programs, as described in Chapter 6.

Chapter 5

Program Synthesis and Extension

*While the design of an alternative construct now seems to be
a reasonably straightforward activity, that of a repetitive
construct requires what I regard as "the invention" of an
invariant relation and a variant function.*

*—Edsger Wybe Dijkstra (Guarded commands,
nondeterminancy and formal derivation of programs)*

5.1. Introduction

Over the years, researchers have tried to gain insight into the
haphazard art of programming. This has led to the development of
"structured programming," which has been defined as "the task of
organizing one's thought in a way that leads, in a reasonable time, to an
understandable expression of a computing task."[1] One of the guidelines
of structured programming is that "one should try to develop a program

---

[1]C. A. R. Hoare, quoted in [Gries74].

and its proof of correctness hand-in-hand."[2] Much has been written on the subject, including [Dijkstra68], [Wirth71], [DahlDijkstraHoare72], [Wirth73], [Wirth74], [ConwayGries75], [Dijkstra76], [Gries81], and others.

As we have seen, it may be necessary to reconstruct pieces of code when modifying a program or instantiating a schema. Our purpose in this chapter is to formalize some of the strategies of structured programming, thereby contributing to the possibility of their automation. The idea is to construct the desired program step by step, beginning with the given input and output specifications. At each step, we use *synthesis rules* to generate code that solves the current goal or to transform the current goal into one or more subgoals. After each rule application, the partially constructed program is correct if its predecessor is, thereby guaranteeing the correctness of the final program. We also describe methods for extending an existing program to achieve additional goals. When modifying a program in this manner, care must be taken to ensure that the original specification continues to be satisfied.

There has been a considerable amount of research into program synthesis. An early approach to automatic programming is [Waldinger69]. Some observations on human synthesis activity may be found in [Standish73]. The system described in [Green76] assumes extensive *a priori* programming knowledge, such as an experienced programmer would have. For surveys of various approaches to automatic program synthesis, see [Biermann76] and [Smith80]. Automatic generation of business applications is treated, for example, in [Goldberg74] and [Prywes77]; numerical programs are discussed in [Brown81].

One of the major hurdles in automatic structured programming lies in the formation of loops. This problem has been variously dealt with. In [BuchananLuckham74] the user was required to supply the skeleton of the loop, and the system filled in details. [Sussman75] describes the HACKER system which creates iterative and recursive loops, but with no guarantee of correctness. [Darlington75], [MannaWaldinger75], [MannaWaldinger80], [Darlington81], and others describe techniques of

---

[2] [Gries74].

recursive loop formation, including the need to sometimes strengthen the original specification for that purpose. In contrast, we concentrate on strategies for iterative loops.

The next section presents an overview of the steps involved in the construction of a simple program and introduces the synthesis rules. In Section 5.3 the rules are employed in the syntheses of several programs. Some of the programs are also extended to achieve additional goals. The last section contains a brief discussion.

5.2. Overview

Program synthesis begins with an initial goal which we present in the form

```
P: begin comment program specification
   assert input specification
   achieve   output specification
             varying output variables
   end
```

Our task is to expand the goal into a segment of code whose execution will always terminate with the relation expressed in the output specification holding between the variables. By

varying *output variables,*

we indicate that only those variables listed may be set by the program; other variables appearing in the specification are input variables. The statement

assert *input specification*

specifies the set of values of the input variables for which the synthesized program is expected to work.

From such a specification, we wish to construct a program of the general form

P: **begin comment** *desired program*
assert *input specification*
purpose *output specification*
   code to achieve specification
assert *output specification*
end

For any input values satisfying the input specification, when control reaches the end of the program, the output specification should be satisfied. The code must also be "primitive," in other words, it may not contain **achieve** statements or nonprimitive operators. It is the task of an automated synthesizer to "compile" given high-level **achieve** statements into lower-level "code." The **purpose** statement is a comment expressing what the indented code that follows was intended to achieve.

It is usually not possible to derive code directly from the initial goal. Rather, at each stage of the construction, some current goal is replaced by one or more new subgoals that are hopefully more readily achievable, and will imply the desired relation, if and when they are achieved. Each step in the synthesis is made by applying a synthesis rule of the general form

achieve $\alpha(\overline{u},\overline{v})$ **varying** $\overline{v}$

purpose $\alpha(\overline{u},\overline{v})$
   code containing new subgoals
assert $\alpha(\overline{u},\overline{v})$

The antecedent of each rule is a program segment containing an **achieve** statement; the consequent is also a program segment. Such a rule is applied to a subgoal matching the antecedent, replacing it with the corresponding consequent, which itself may contain new goals. Each rule application preserves correctness, i.e. satisfying the new goals yields a correct program satisfying the current goal. Thus, the final program is guaranteed to satisfy the original specification. Our synthesis rules are labeled $<a>$, $<b>$, etc. and are listed together in Appendix 3.

In the rest of this section, we use the simple and familiar "greatest common divisor" problem to illustrate aspects of the methodology. The goal is

P_7: **begin comment** *gcd specification*
  assert $a \in \mathbf{N}$, $b \in \mathbf{N}+1$
  achieve $z = gcd(a,b)$ **varying** z
  end

where $\mathbf{N}$ is the set of natural numbers and $\mathbf{N}+1$ the set of positive integers. We are to construct a program that sets the variable z to the greatest common divisor (gcd) of two nonnegative integers a and b. To illustrate the various rules, we give three alternative syntheses, as depicted in Figure 2. The next section presents several more involved examples.

Were the function gcd a primitive of the target language, then the above goal could be achieved by a simple assignment statement,

$$z := gcd(a,b).$$

But with no primitive gcd function available, it must be achieved in stages. Nor can we use the definition

$$\mathbf{fact} \ gcd(u,v) = \max_{w \in \mathbf{N}}(w \mid u \wedge w \mid v)$$

(where the predicate $w \mid u$ means that w divides u evenly) to assign

$$z := \max_{w \in \mathbf{N}}(w \mid a \wedge w \mid b),$$

since the range $w \in \mathbf{N}$ of the max operator is infinite. Note also that were it not specified that only z may be varied, the goal could be achieved by the assignment

$$(z,a) := (b,0),$$

since $b = gcd(0,b)$.

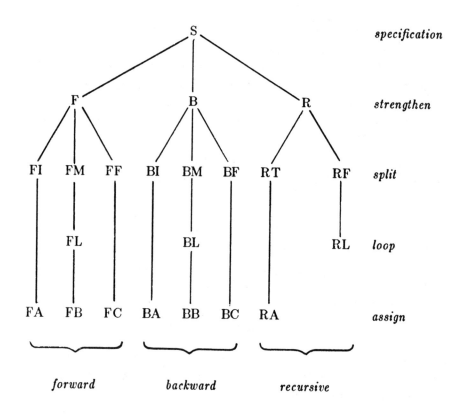

Figure 3. *Alternative syntheses of a simple program.*

5.2.1. Strengthening

With no immediate way to solve the goal

$$\textbf{achieve } z = gcd(a,b) \textbf{ varying } z, \tag{S}$$

the first step in a synthesis might be to introduce additional program variables, the values of which may be manipulated by the program so as to achieve the goal in stages. For example, the goal may be replaced by the conjunction of the following two subgoals:

$$\textbf{achieve } z = gcd(s,t), \; gcd(s,t) = gcd(a,b) \textbf{ varying } z,s,t. \tag{F}$$

The first subgoal $z = gcd(s,t)$ states that z should have as its value the gcd of two new variables, s and t, whose values are to be set by the synthesized program. If the second subgoal $gcd(s,t) = gcd(a,b)$ holds as well, then the original goal $z = gcd(a,b)$ will also be satisfied.

The above step is an instance of the following rule, used to replace one goal by another sufficient subgoal:

$<$s$>$ *strengthening rule*
assert γ **achieve** $\alpha(\overline{u})$ **varying** $\overline{u}$ **fact** $\alpha(\overline{u})$ **when** $\beta(\overline{u},\overline{v})$, γ
assert γ **purpose** $\alpha(\overline{u})$ **achieve** $\beta(\overline{u},\overline{v})$ **varying** $\overline{u},\overline{v}$ **assert** $\alpha(\overline{u})$

where γ does not contain the variables $\overline{u}$. The antecedent of this rule includes a **fact** that must hold for the rule to be applicable. Thus, this rule states that if achieving β will imply the desired relation α, given that γ holds, then the goal

$$\textbf{achieve } \alpha(\overline{u}) \textbf{ varying } \overline{u}$$

may be replaced by the "stronger" subgoal

$$\textbf{achieve } \beta(\overline{u},\overline{v}) \textbf{ varying } \overline{u},\overline{v}.$$

The goal β may introduce new variables $\overline{v}$, in which case β must imply α for any values of $\overline{v}$, i.e.

$$(\forall \overline{u}, \overline{v})\ [\gamma \wedge \beta(\overline{u},\overline{v}) \supset \alpha(\overline{u})].$$

In particular, α will hold for the values of $\overline{v}$ set by the code generated for this subgoal; what, specifically, the final values of $\overline{v}$ are is unimportant, as they are not output variables. A special case of this rule is the replacement of a goal by a logically equivalent (but simpler) goal.

Some generally useful transformations for introducing new variables are expressed by the following facts:

fact $p(\overline{u})$ **when** $p(\overline{v})$, $p(\overline{v}) \supset p(\overline{u})$
fact $p(\overline{u})$ **when** $p(\overline{v})$, $\overline{u}=\overline{v}$
fact $p(f(\overline{u}))$ **when** $p(f(\overline{v}))$, $f(\overline{u})=f(\overline{v})$
fact $p(f(\overline{u}))$ **when** $p(v)$, $v=f(\overline{u})$.

For example, the goal

achieve $z=gcd(a,b)$ **varying** z　　　　　(S)

may be transformed into

achieve $z=gcd(s,t)$, $(s,t)=(a,b)$ **varying** z,s,t,　　　(B)

as well as to

achieve $z=gcd(s,t)$, $gcd(s,t)=gcd(a,b)$ **varying** z,s,t.　　(F)

In either case, $z=gcd(a,b)$ is implied, regardless of the values of s and t.

The purpose of this rule is to *simplify* goals. Thus, the choice of which fact to apply when is crucial. Facts may be very general, as the four listed above, or specific to the problem-domain in question. [Dijkstra76] and [Gries81] contain suggestions for transforming, and strengthening, goals.

5.2.2. Splitting

Suppose we have a goal of the form

achieve β, γ **varying** $\overline{v}$.

We would like to simplify such a conjunctive goal by splitting it into consecutive subgoals, say, first achieving β and then γ:

> **purpose** β, γ
>   **achieve** β **varying** $\overline{v}$
>   **achieve** γ **varying** $\overline{v}$
> **assert** β, γ.

For example, to

$$\text{\textbf{achieve}} \ z = gcd(s,t), \ gcd(s,t) = gcd(a,b) \ \textbf{varying} \ z,s,t, \quad (\text{F})$$

we might choose to first achieve $gcd(s,t) = gcd(a,b)$ and then $z = gcd(s,t)$, each of which is simpler than the conjunctive goal. Unfortunately, things are not as easy as that. The problem is that in achieving the second subgoal γ, the relation β that has been achieved by the first subgoal may be unwittingly destroyed. To ensure that achieving the second subgoal will not undo what was accomplished by the first subgoal, we consider three strategies.

5.2.2.1. Disjoint Goals

If in achieving the second subgoal γ, the values of the variables set by the first subgoal β can be left untouched, then the two subgoals may be solved independently. Thus, one approach to conjunctive goals is to form consecutive subgoals using the rule

$<\text{d}>$ *disjoint goal rule*
achieve $\beta(\overline{v})$, $\gamma(\overline{u},\overline{v})$ **varying** $\overline{u},\overline{v}$

purpose $\beta(\overline{v})$, $\gamma(\overline{u},\overline{v})$
   achieve $\beta(\overline{v})$ **varying** $\overline{v}$
   achieve $\gamma(\overline{u},\overline{v})$ **varying** $\overline{u}$
 assert $\beta(\overline{v})$, $\gamma(\overline{u},\overline{v})$

For example, the two conjuncts of the goal

$$\text{\textbf{achieve}} \ z = gcd(s,t), \ gcd(s,t) = gcd(a,b) \ \textbf{varying} \ z,s,t \quad (\text{F})$$

do not both contain all three variables. We can therefore split it into

> **purpose** $z=gcd(s,t)$, $gcd(s,t)=gcd(a,b)$
>   **achieve** $gcd(s,t)=gcd(a,b)$ **varying** s,t
>   **achieve** $z=gcd(s,t)$ **varying** z
> **assert** $z=gcd(s,t)$, $gcd(s,t)=gcd(a,b)$.

This strategy, unfortunately, gets us no place in this case, since achieving $z=gcd(s,t)$ by varying only z is no easier than the original goal of achieving $z=gcd(a,b)$ for input values a and b. For this conjunctive goal, a different splitting strategy is required.

5.2.2.2. Protection

Another splitting strategy is to insist that after each stage executed in achieving the second goal γ, the first goal β remain true for the current values of the variables. This idea is expressed in the rule

$<$p$>$ *protection rule*
achieve $\beta(\overline{v})$, $\gamma(\overline{u},\overline{v})$ **varying** $\overline{u},\overline{v}$

purpose $\beta(\overline{v})$, $\gamma(\overline{u},\overline{v})$
    achieve $\beta(\overline{v})$ **varying** $\overline{v}$
    achieve $\gamma(\overline{u},\overline{v})$ **protecting** $\beta(\overline{v})$
                          varying $\overline{u},\overline{v}$
 assert $\beta(\overline{v})$, $\gamma(\overline{u},\overline{v})$

In the *disjoint goal rule,* the relation β was protected by insisting that its variables do not change value. This is a very stringent restriction. In general we need only require that $\beta(\overline{v}')$ imply $\beta(\overline{v})$, where $\overline{v}'$ denotes the values of the variables prior to the code for achieving γ. This is the meaning of the **protecting** clause. One means of protecting an achieved goal is the formation of a loop, as in the *forward loop rule* below. For example, the goal

$$\text{\textbf{achieve} } z=gcd(s,t),\ gcd(s,t)=gcd(a,b) \text{ \textbf{varying} } z,s,t \qquad (F)$$

may be broken into

$$\textbf{achieve } gcd(s,t) = gcd(a,b) \textbf{ varying } s,t \qquad\qquad \text{(FI)}$$

$$\textbf{achieve } z = gcd(s,t) \qquad \textbf{protecting } gcd(s,t) = gcd(a,b) \quad \text{(FMF)}$$
$$\textbf{varying } z,s,t.$$

That is, first set s and t in the easiest manner that establishes $gcd(s,t) = gcd(a,b)$. Then, try to achieve $z = gcd(s,t)$ by varying z, s, and t, at the same time maintaining $gcd(s,t) = gcd(a,b)$.

5.2.2.3. Preservation

Suppose that some of the variables in the conjunctive goals β and γ are not output variables, but, rather, were introduced to facilitate achieving some purpose α. In that case, the final values of these variables $\overline{v}$ are unimportant; as long as β and γ hold for *some* set of values of $\overline{v}$, the real goal α will be satisfied. As we saw, the *protection rule* achieves both β and γ for the final values of $\overline{v}$; while in the *disjoint rule*, the values of $\overline{v}$ are the same after achieving γ as after achieving β. A third possibility is to achieve γ for those values of $\overline{v}$ for which β was found to hold, though the values of $\overline{v}$ may change in the process. The only requirement is that achieving γ for the new values of $\overline{v}$ preserve the truth of γ for the previous values. Equivalently, if γ was the goal for the old $\overline{v}$, then γ remains the goal after this stage—for the new $\overline{v}$. Thus, by achieving γ, we end up with β and γ both holding for the *old* values of $\overline{v}$. The appropriate rule is

$<$v$>$ *preservation rule*
purpose $\alpha(\overline{u})$ **achieve** $\beta(\overline{v})$, $\gamma(\overline{u},\overline{v})$ **varying** $\overline{u},\overline{v}$

purpose $\alpha(\overline{u})$
     achieve $\beta(\overline{v})$ **varying** $\overline{v}$
     achieve $\gamma(\overline{u},\overline{v})$ **preserving** $\gamma(\overline{u},\overline{v})$ **for** $\overline{v}$
                             varying $\overline{u},\overline{v}$
 assert $\alpha(\overline{u})$

The **preserving** clause means that $\gamma(\overline{u},\overline{v})$ must imply $\gamma(\overline{u},\overline{v}')$, *for* the values $\overline{v}'$ attained in achieving β. This rule can lead to a loop in which γ remains the goal (see the *backward loop rule* below).

For example, the goal

$$\textbf{achieve } z=gcd(s,t),\ (s,t)=(a,b)\ \textbf{varying } z,s,t \qquad \text{(B)}$$

may be broken into two subgoals,

$$\textbf{achieve } (s,t)=(a,b)\ \textbf{varying } s,t \qquad \text{(BI)}$$
$$\textbf{achieve } z=gcd(s,t) \quad \textbf{preserving } z=gcd(s,t)\ \textbf{for } s,t \quad \text{(BMF)}$$
$$\textbf{varying } z,s,t,$$

since our ultimate $z=gcd(a,b)$ does not contain s or t. Matching the second subgoal with the domain-specific knowledge

$$\textbf{fact } gcd(0,u)=u\ \textbf{when } u\in N+1$$

(if u is positive, then the gcd of u and 0 is u), suggests that $z=gcd(s,t)$ if $s=0$ and $z=t\in N+1$.[3] Thus, it may be strengthened to

$$\textbf{achieve } s=0,\ t\in N+1,\ z=t \quad \textbf{preserving } z=gcd(s,t)\ \textbf{for } s,t$$
$$\textbf{varying } z,s,t.$$

Now one way to split this conjunctive goal is into the two disjoint subgoals

$$\textbf{achieve } s=0,\ t\in N+1 \quad \textbf{preserving } z=gcd(s,t)\ \textbf{for } s,t \quad \text{(BM)}$$
$$\textbf{varying } s,t$$
$$\textbf{achieve } z=t \quad \textbf{preserving } z=gcd(s,t)\ \textbf{for } s,t \quad \text{(BF)}$$
$$\textbf{varying } z.$$

This means that we reset s to 0 and t to some positive integer, in such a way as to ensure that, if $z=gcd(s,t)$ for these new values of s and t, then $z=gcd(s',t')$ for the previous values $s'=a$ and $t'=b$, as well. At that point, we set z to equal t to achieve the goal $z=gcd(s,t)$.

To summarize the difference between the last two rules, we may say that the *protection rule* maintains goals *already* achieved, while the *preservation rule* maintains goals *to be* achieved. The achievement of

---

[3]Realistically, such high-level facts about the problem domain are not always readily available. That would necessitate, in our case, first deriving the fact that $gcd(0,u)=u$ from more basic facts about divisibility.

conjunctive goals is the topic of [Waldinger77]; protection mechanisms are used for this purpose by [Sussman75]; [Sacerdoti75] addresses their nonlinear nature.

5.2.3. Assignments

Assignment statements are formed by the following rule:

$<$a$>$ *assignment rule*
achieve $y_1 = f_1(\overline{x}),\ y_2 = f_2(\overline{x}),\ \cdots,\ y_n = f_n(\overline{x})$ $\qquad$ **varying** $y_1, y_2, \cdots, y_n, \cdots$
purpose $y_1 = f_1(\overline{x}),\ y_2 = f_2(\overline{x}),\ \cdots,\ y_n = f_n(\overline{x})$ $\quad(y_1, y_2, \cdots, y_n) := (f_1(\overline{x}), f_2(\overline{x}), \cdots, f_n(\overline{x}))$ **assert** $y_1 = f_1(\overline{x}),\ y_2 = f_2(\overline{x}),\ \cdots,\ y_n = f_n(\overline{x})$

where the variables y_1, y_2, ... , y_n do not appear in $\overline{x}$, and the expressions f_1, f_2, ... , f_n are composed of only primitive operations. This rule suggests that one first attempt to isolate variables on one side of an equality. For example, a goal of the form

$$\textbf{achieve } g(y) = f(x) \textbf{ varying } y$$

can be transformed into

$$\textbf{achieve } y = g^-(f(x)) \textbf{ varying } y,$$

where g^- is the inverse of the function g, i.e. $g(g^-(x)) = x$ for all x.

The first and last subgoals in the program segment

achieve $(s,t) = (a,b)$ **varying** s,t		(BI)
achieve $s = 0,\ t \in N+1$	**preserving** $z = gcd(s,t)$ **for** s,t **varying** s,t	(BM)
achieve $z = t$	**preserving** $z = gcd(s,t)$ **for** s,t **varying** z	(BF)

may be achieved by assignments. The first gives rise to the statement

$$(s,t) := (a,b). \qquad\qquad \text{(BA)}$$

By assigning

$$z := t, \tag{BC}$$

the third subgoal $z = t$ is achieved, as is the preserved goal $z = gcd(s,t)$, since at that point $s = 0$.

Similarly, after applying the *protection rule*, we were left with the fragment

> **achieve** $gcd(s,t) = gcd(a,b)$ **varying** s,t (FI)
> **achieve** $z = gcd(s,t)$ **protecting** $gcd(s,t) = gcd(a,b)$ (FMF)
>   **varying** z,s,t.

The second of the two goals may be replaced by the stronger

> **achieve** $z = t$, $t \in N+1$, $s = 0$ **protecting** $gcd(s,t) = gcd(a,b)$
>   **varying** z,s,t,

and then split into disjoint goals. That gives

> **achieve** $gcd(s,t) = gcd(a,b)$ **varying** s,t (FI)
> **achieve** $s = 0$, $t \in N+1$ **protecting** $gcd(s,t) = gcd(a,b)$ (FM)
>   **varying** s,t
> **achieve** $z = t$ **protecting** $gcd(s,t) = gcd(a,b)$ (FF)
>   **varying** z.

Notice that we have allowed the second subgoal to vary s and t, but not z, while the third may vary only z, leaving the values of s and t unchanged. The first subgoal may be strengthened to

> **achieve** $(s,t) = (a,b)$ **varying** s,t,

and then achieved by the assignment

$$(s,t) := (a,b). \tag{FA}$$

The third subgoal may be achieved by assigning

$$z := t. \tag{FC}$$

5.2.4. Conditionals

Conditional statements are formed in the following manner:

$<c>$ *conditional rule*
purpose α **achieve** β, γ **varying** $\overline{v}$

purpose α
    if β **then**   **achieve** γ **protecting** β
                               varying $\overline{v}$
         else   **assert** $\sim\beta$
                 achieve α **varying** $\overline{v}$ **fi**
 assert β, γ

provided that the relation β is computable, i.e. when β is composed of primitive functions and predicates. The **else** case attempts to achieve the original goal α, extracted from the **purpose** statement, since it is in general weaker than the derived goal β and γ. In other words, one way of achieving β is to test if it holds: when it does, protect that relation while achieving the remainder; when it does not, try to use that fact while achieving the original goal. For example, the original goal

$$\textbf{achieve } z = gcd(a,b) \textbf{ varying } z \tag{S}$$

could be strengthened to

$$\begin{aligned} &\textbf{purpose } z = gcd(a,b)\\ &\qquad\textbf{achieve } a = 0,\ z = b \textbf{ varying } z \end{aligned} \tag{R}$$

using the

$$\textbf{fact } gcd(0,u) = u \textbf{ when } u \in N + 1.$$

With this goal, one might test if the first conjunct already holds:

$$\begin{aligned} &\textbf{purpose } z = gcd(a,b)\\ &\quad\textbf{if } a = 0 \quad \textbf{then} \quad \textbf{achieve } z = b \textbf{ varying } z \quad\quad\text{(RT)}\\ &\qquad\qquad\quad \textbf{else} \quad \textbf{assert } a \neq 0\\ &\qquad\qquad\qquad\quad \textbf{achieve } z = gcd(a,b) \textbf{ varying } z \quad\textbf{fi.} \quad \text{(RF)} \end{aligned}$$

That leaves two subgoals, for the cases when $a = 0$ and $a \neq 0$. There is no need in this case to protect the condition $a = 0$ since a cannot be varied.

Conditional-formation techniques are the subject of investigation in [LuckhamBuchanan74] and [Warren76].

5.2.5. Loops

For the "forward" synthesis of the gcd program, we still need to achieve the conjunctive goal

achieve $s=0$, $t \in N+1$ **protecting** $gcd(s,t)=gcd(a,b)$ (FM)
varying s,t.

To achieve $s=0$, we cannot simply assign $s := 0$, since this will undo the protected relation; on the other hand, we must vary the value of s, since the current value a of s is not necessarily 0.

The loop rules allow such a goal to be achieved step by step. We present rules for forming iterative loops from conjunctive goals and for guaranteeing their termination. For completeness, we also include a recursion-formation rule.

5.2.5.1. Forward Iterative Loops

To achieve $s=0$ while protecting $gcd(s,t)=gcd(a,b)$, we attempt a loop of the form

purpose $s=0$, $gcd(s,t)=gcd(a,b)$
  loop L_7: **assert** $gcd(s,t)=gcd(a,b)$
     until $s=0$
     approach $s=0$ **protecting** $gcd(s,t)=gcd(a,b)$ (FL)
                varying s,t
  repeat
assert $s=0$, $gcd(s,t)=gcd(a,b)$.

The statement

L_7: **assert** $gcd(s,t)=gcd(a,b)$

declares that the protected relation is the *invariant assertion* of the loop.[4] The new goal

> **approach** $s=0$ **protecting** $gcd(s,t)=gcd(a,b)$ (FL)
> **varying** s,t,

as we will see, can be achieved by decreasing the value of s. The invariant must be protected in the process.

In general, given a goal of the form

> **achieve** β **protecting** α
>   **varying** $\overline{v}$

(preceding code has achieved α, and we wish to keep α true while achieving β), the following rule will generate an iterative loop:

$<$f$>$ *forward iterative loop rule*
achieve β **protecting** α **varying** $\overline{v}$

purpose β, α
   loop L: **assert** α
        until β
        approach β **protecting** α
              varying $\overline{v}$
        repeat
 assert β, α

This rule is useful only when the exit test β is primitive. The statement

$$L: \textbf{assert } \alpha$$

is the invariant assertion of the loop. It is invariant, since it is given that α is true when the loop is first entered (and only needs to be protected), and the loop-body subgoal

> **approach** β **protecting** α
>   **varying** $\overline{v}$

will ensure that α remains true after each iteration. The loop will

---

[4]Recall that invariants are associated with specific points in a program segment, and express that part of the goal that has already been computed whenever execution reaches the corresponding point.

terminate when the exit condition β becomes true; at that point both the invariant α and the test β must hold. In order to guarantee that loop execution will indeed terminate, we must make definite progress towards β; this is the meaning of "approach."

5.2.5.2. Backward Iterative Loops

Consider the program segment

$$(s,t) := (a,b) \tag{BA}$$

achieve $s=0$ **preserving** $z=gcd(s,t)$ **for** s,t (BM)

                        varying s,t

$$z := t. \tag{BC}$$

Recall that the purpose in having introduced the variables s and t and having assigned $(s,t) := (a,b)$ was to enable us to compute $z=gcd(s,t)$ rather than $z=gcd(a,b)$. We must make sure that, though the value of s is to be changed, it will still suffice to achieve $z=gcd(s,t)$. The relation $z=gcd(s,t)$ is then called an *invariant purpose;* it expresses the ultimate goal of the remainder of the computation: the final value of z should be the *gcd* of the current values of s and t.[5]

For a relation to be an invariant purpose of a loop, it must be the goal upon entering the loop, and, assuming that it is the goal before executing the loop body, it must be the goal after. For such a case where we wish to achieve a relation β while preserving a ultimate goal α, we have the rule

---

[5] The *invariant purpose* is the assertion used for "subgoal induction;" see [Manna71], [BasuMisra75], and [MorrisWegbreit77].

<b> *backward iterative loop rule*

---

achieve $\beta(\overline{u},\overline{v})$ **preserving** $\alpha(\overline{u},\overline{v})$ **for** $\overline{v}$
            varying $\overline{u},\overline{v}$

---

purpose $\beta(\overline{u},\overline{v})$
    loop L : **purpose** $\alpha(\overline{u},\overline{v})$ **for** $\overline{v}$
               until $\beta(\overline{u},\overline{v})$
               approach $\beta(\overline{u},\overline{v})$ **preserving** $\alpha(\overline{u},\overline{v})$ **for** $\overline{v}$
                         varying $\overline{u},\overline{v}$
               repeat
assert $\beta(\overline{u},\overline{v})$

As with the forward loop, this rule is useful only if β is computable. The purpose of the loop is to achieve the exit relation β while preserving the ultimate purpose α. The loop will terminate when the exit condition β becomes true; at that point, the fact that β holds may be used to help achieve the "larger" purpose α. The statement

$$\textbf{purpose } \alpha(\overline{u},\overline{v}) \textbf{ for } \overline{v}$$

contains the invariant purpose of the loop and states that whenever execution reaches the beginning of the loop, what remains to be computed is α for the current values of the variables $\overline{v}$. Upon exiting the loop, the goal is still α, and α is the goal whenever the loop-body subgoal

$$\textbf{approach } \beta(\overline{u},\overline{v}) \quad \textbf{preserving } \alpha(\overline{u},\overline{v}) \textbf{ for } \overline{v}$$
$$\textbf{varying } \overline{u},\overline{v}$$

is executed.

Letting α be $z=gcd(s,t)$ and β be $s=0$ in the above rule, the goal

$$\textbf{achieve } s=0 \quad\quad \textbf{preserving } z=gcd(s,t) \textbf{ for } s,t \quad\quad \text{(BM)}$$
$$\textbf{varying } s,t$$

may be transformed into the loop

> **purpose** $s = 0$
>> **loop** L_7: **purpose** $z = gcd(s,t)$ **for** s, t
>> **until** $s = 0$
>> **approach** $s = 0$ **preserving** $z = gcd(s,t)$ **for** s, t (BL)
>>> **varying** s, t
>> **repeat**
> **assert** $s = 0$.

Within the loop, the purpose $z = gcd(s,t)$ must be maintained while making progress towards $s = 0$.

5.2.5.3. Termination

In both the forward and backward versions of the loop, the statement

$$\textbf{approach } s = 0 \cdots \textbf{ varying } s, t \qquad \text{(FL,BL)}$$

expresses the desire to make definite progress towards the goal $s = 0$ with each loop iteration. Since s is set to the nonnegative integer a before entering the loop, it follows that the loop decreases the value of s. Thus, we can ensure loop termination by monotonically decreasing the integer s, while s remains nonnegative; the loop-body subgoal would be

$$\textbf{achieve } s \in N, \; s < s' \textbf{ varying } s, t,$$

where s' denotes the value of s prior to this statement.

In general, assume that we are given a loop-body subgoal of the form

$$\textbf{approach } \beta \textbf{ varying } \overline{v}.$$

Clearly in order to make progress towards β, one of the variables $\overline{v}$ must be changed, i.e. $\overline{v} \neq \overline{v}'$. This is not however sufficient to ensure that β will ever be attained. What we need is the notion of well-founded set: a *well-founded set* $(W, >)$ consists of a set of elements W and an ordering $>$ defined on the elements, such that there can be no infinite descending sequences of elements $w_1 > w_2 > \cdots$. So, if throughout execution of the loop we keep $\overline{v} \in W$, for some well-founded set $(W, >)$, and insist that with each iteration $\overline{v}$ is reduced in that ordering, i.e. $\overline{v}' > \overline{v}$, then termination is guaranteed.[6] That gives a loop-body subgoal

achieve $\bar{v} < \bar{v}'$, $\bar{v} \in W$ **varying** $\bar{v}$.

If the value of $\bar{v}$ is going to decrease in the well-founded set, then in particular, we must have $\bar{v}_B \geq \bar{v}_E$, where $\bar{v}_B$ denotes the value of $\bar{v}$ upon entering the loop and $\bar{v}_E$ denotes the value upon exiting.[7] That gives, for forward loops, the rule

$<\text{t}>$ *forward termination rule*
assert $\bar{v}_B \in W$, $\bar{v}_E \in W$, $\bar{v}_B \geq \bar{v}_E$ **approach** $\beta(\bar{v})$ **protecting** $\alpha(\bar{v})$ **varying** $\bar{v}$
assert $\sim\beta(\bar{v})$ **achieve** $\bar{v} < \bar{v}'$ **protecting** $\alpha(\bar{v})$, $\bar{v} \in W$ **varying** $\bar{v}$

and similarly for backward loops

$<\text{m}>$ *backward termination rule*
assert $\bar{v}_B \in W$, $\bar{v}_E \in W$, $\bar{v}_B \geq \bar{v}_E$ **approach** $\beta(\bar{v})$ **preserving** $\alpha(\bar{v})$ **for** $\bar{v}$ **varying** $\bar{v}$
assert $\sim\beta(\bar{v})$ **achieve** $\bar{v} < \bar{v}'$ **preserving** $\alpha(\bar{v})$, $\bar{v} \in W$ **for** $\bar{v}$ **varying** $\bar{v}$

where $(W, >)$ is any well-founded set.

The well-founded set most commonly used for termination proofs is the set **N** of nonnegative integers under the $>$ ordering. When dealing with more than one variable, the lexicographic ordering on n-tuples is useful. For example, to use a lexicographic ordering on the values of two variables x and y, we first look for well-founded sets W_1 and W_2 such that $x \in W_1$ and $y \in W_2$. Then we consider the pair (x, y) and require

---

[6]See [Floyd67].

[7]To determine this, one may use whatever facts are known about $\bar{v}_B$ and $\bar{v}_E$, including $\alpha(\bar{v}_B)$, $\alpha(\bar{v}_E)$, and $\beta(\bar{v}_E)$.

achieve $(x,y) <_2 (x',y')$ **protecting** $\alpha(x,y)$, $x \in W_1$, $y \in W_2$
varying x,y

where $>_2$ is the lexicographic ordering on pairs, i.e. we must achieve $x < x'$ or else $x' = x$ and $y < y'$ while protecting the relation $\alpha(x,y)$ as well as $x \in W_1$ and $y \in W_2$.

Another, often useful, tactic for handling several variables is based on an assumption of monotonicity for each variable. Determining, for some variable x, that the initial value x_B is greater than the final value x_E suggests that x decrease monotonically, i.e. $x \le x'$. In that case, we may also

assert $x_E \le x \le x_B$

within the loop (if we can determine the values of x_B and x_E). Given two monotonic variables x and y in well-founded sets W_1 and W_2, respectively, termination may be ensured by requiring

achieve $x \le x'$, $y \le y'$, $x < x' \lor y < y'$
protecting $\alpha(x,y)$, $x \in W_1$, $y \in W_2$, $x \ge x_E$, $y \ge y_E$
varying x,y.

In other words, each iteration reduces the value of one of the variables, without increasing the other.

Returning to our *gcd* program, the conjunct $t \in N+1$ of the current subgoal is true upon entering the loop, when $t = b \in N+1$, and must be kept true by the loop. Within the loop, we also wish to protect the invariant assertion $gcd(s,t) = gcd(a,b)$ or the invariant purpose $z = gcd(s,t)$. If $gcd(s',t') = gcd(a,b)$ holds for the prior values of s and t, then $gcd(s,t) = gcd(a,b)$ will hold for the new values, provided that $gcd(s,t) = gcd(s',t')$. This relation also maintains the goal $z = gcd(s,t)$: if $gcd(s,t) = gcd(s',t')$ and the goal $z = gcd(s,t)$ can be achieved, then $z = gcd(s',t')$ will be achieved as well.

The complete loop-body subgoal is

> **achieve** $s \in N$, $s < s'$, $t \in N+1$, $gcd(s,t)=gcd(s',t')$
> **varying** s,t.   (FL,BL)

One way to solve this goal is by matching the conjunct $gcd(s,t)=gcd(s',t')$ with the information about the domain expressed in the

> **fact** $gcd(u,v+w\cdot u)=gcd(u,v)$ **when** $w \in Z$

(where Z is the set of all integers). This yields a new goal

> **achieve** $s \in N$, $s < s'$, $t \in N+1$, $gcd(s,t)=gcd(s',t'+w\cdot s')$, $w \in Z$
> **varying** s,t,w.

Taking into consideration the commutativity of gcd, i.e.

> **fact** $gcd(u,v)=gcd(v,u)$,

and the

> **fact** $p(\overline{u})$ **when** $p(\overline{v})$, $\overline{u}=\overline{v}$,

suggests as one possible goal

> **achieve** $s \in N$, $s < s'$, $t \in N+1$,
> $t=s'$, $s=t'+w\cdot s'$, $w \in Z$
> **varying** s,t,w,

or (substituting for s and t)

> **achieve** $t'+w\cdot s' \in N$, $t'+w\cdot s' < s'$, $s' \in N+1$,
> $t=s'$, $s=t'+w\cdot s'$, $w \in Z$
> **varying** s,t,w.

For the loop to have been continued, the exit test $s=0$ must have been false, from which it follows that $s' \in N+1$. The two conjuncts $t'+w\cdot s' \in N$ and $t'+w\cdot s' < s'$ together imply $t'/s'-1 \leq -w < t'/s'$. With the conjunct $w \in Z$, that forces w to be $\lfloor t'/s' \rfloor$, since

> **fact** $v=\lfloor u \rfloor$ **is** $v \leq u$, $u < v+1$, $v \in Z$.

Thus, we are left with the goal

> **achieve** $t=s'$, $s=t'-\lfloor t'/s' \rfloor \cdot s'$ **varying** s,t,

suggesting the multiple assignment

$$(s,t) := (rem(t,s),s), \qquad \text{(FB,BB)}$$

since, by definition,

fact $rem(u,v) = u - \lfloor u/v \rfloor \cdot v$ **when** $u \in \mathbf{N}$, $v \in \mathbf{N}+1$.

Since for each step in the construction, achieving the new subgoals satisfies the previous goal, the final program is guaranteed to achieve the initial specification. Both the "forward" and "backward" methods have resulted in the same program

---

P_7: **begin comment** *gcd program*
  B_7: **assert** $a \in \mathbf{N}$, $b \in \mathbf{N}+1$
  purpose $z = gcd(a,b)$
    purpose $gcd(s,t) = gcd(a,b)$
    $(s,t) := (a,b)$
    assert $gcd(s,t) = gcd(a,b)$
    purpose $z = gcd(s,t)$
      loop L_7: **assert** $gcd(s,t) = gcd(a,b)$
              purpose $z = gcd(s,t)$ **for** s,t
      until $s = 0$
      $(s,t) := (rem(t,s),s)$
      repeat
    $z := t$
  E_7: **assert** $z = gcd(a,b)$
end

5.2.5.4. Recursive Loops

Suppose the current goal in a program derivation is

achieve $\alpha(\overline{v})$ **varying** $\overline{v}$,

while the code that is being synthesized—call it P—has the similar

purpose $\alpha(\overline{u})$.

This situation suggests forming a recursive loop, by inserting a recursive call $P(\overline{v})$ to achieve the current goal, provided, however, that the input assertion for P is satisfied by the arguments $\overline{v}$. Furthermore, to guarantee that the recursion will not continue forever, it must also be that $\bar{v} < \bar{u}$ in some well-founded ordering $(W, >)$. If these requirements are satisfied, then a recursive call to P will accomplish the desired goal. The following rule is used:

$<r>$ *recursive loop rule*

assert $\gamma(\bar{u})$, $\bar{u} \in W$
 purpose $\alpha(\bar{u})$
 $\cdots$
      assert $\gamma(\bar{v})$, $\bar{v} \in W$, $\bar{v} < \bar{u}$
      achieve $\alpha(\bar{v})$ **varying** $\bar{v}$
 $\cdots$
 assert $\alpha(\bar{u})$

assert $\gamma(\bar{u})$, $\bar{u} \in W$
 $P(\bar{u})$: **procedure**
      purpose $\alpha(\bar{u})$
 $\cdots$
      assert $\gamma(\bar{v})$, $\bar{v} \in W$, $\bar{v} < \bar{u}$
      purpose $\alpha(\bar{v})$
        $P(\bar{v})$
      assert $\alpha(\bar{v})$
 $\cdots$
      assert $\alpha(\bar{u})$
 end

where the code bracketed by **procedure** and **end** is executed for the values of $\bar{u}$ and recursively repeated for values $\bar{v}$.

For example, after applying the *conditional rule*, we were left with the goal

purpose $z = gcd(a,b)$
     if $a = 0$ **then**   **achieve** $z = b$ **varying** z (RT)
              else   **assert** $a \neq 0$
                          achieve $z = gcd(a,b)$ **varying** z **fi** (RF)

which can be reduced to

purpose $z = gcd(a,b)$
    if $a = 0$ **then** $z := b$ (RA)
              else   **assert** $rem(b,a) \in \mathbf{N},$
                                   $a \in \mathbf{N} + 1, \ rem(b,a) < a$
                      achieve $z = gcd(rem(b,a),a)$ **varying** z **fi** (RL)

with reasoning similar to that used for the iterative loop body. The *recursive loop rule* would then produce the program

---

$P_7' \ (a,b,z)$: **procedure comment** *recursive gcd program*
  B_7': **assert** $a \in \mathbf{N}, \ b \in \mathbf{N} + 1$
  if $a = 0$ **then** $z := b$
              else   **assert** $rem(b,a) \in \mathbf{N}, \ a \in \mathbf{N} + 1, \ rem(b,a) < a$
                    $P_7' \ (rem(b,a),a,z)$ **fi**
  E_7': **assert** $z = gcd(a,b)$
  end

The use of invariants for the automatic construction of iterative loops is discussed by [Duran75]. [BasuMisra75] gives criteria for a loop to be formable directly from the program specification. Recursion-formation techniques are discussed in detail by [Siklossy74], [Darling-ton75], [MannaWaldinger79], and others.

5.2.6. Extension

There are two basic ways of extending a given program to achieve additional relations. One is to append code at the end of the program with the purpose of achieving an additional goal, while making sure that the relations already achieved by the program remain intact. The second method is to achieve the added relations at the outset and modify the program to ensure that they are maintained true until the end of execution.

This second method has also been used for local optimization: if a program contains an expression that is relatively difficult to compute, but must be recomputed for each loop iteration, it may be possible to introduce a program variable that will invariantly contain the value of the complex expression, and for which there is a relatively simple way of deriving the new value of the variable from the old value. This new variable must be updated whenever the value of a variable in the expression is changed, and may be substituted for that expression wherever is occurs in the program text.[8]

Consider, for example, the real-division program[9]

P_1: **begin comment** *real–division program*
   B_1: **assert** $0 \leq c < d$, $e > 0$
   $(q,s) := (0,1)$
   loop L_1: **assert** $d \cdot q \leq c$, $c < d \cdot (q + s)$
     until $s \leq e$
     $s := s/2$
     if $d \cdot (q + s) \leq c$ **then** $q := q + s$ **fi**
     repeat
   E_1: **assert** $|c/d - q| < e$
   end

Since the conditional test $d \cdot (q + s) \leq c$ contains the only instance of multiplication in the program, it is a natural candidate for simplification. Testing for $d \cdot q + d \cdot s \leq c$ would be worse, but if we introduce two new variables qq and ss and maintain $d \cdot q$ and $d \cdot s$ as their respective values, then the test can be replaced by the simpler $qq + ss \leq c$. The crucial question, then, is whether qq and ss can themselves be maintained without recourse to multiplication.

Every time that q or s is assigned to, the new variables qq and ss, whose values depend on q and s, respectively, must be updated to reflect the change. When q is given the initial value 0, the new variable qq should have the value $d \cdot 0 = 0$. Similarly, when s is initialized to 1, the variable ss should be set to $d \cdot 1 = d$. When, within the loop, s is halved, the new value of ss must also be halved, so that it remains d times the value of s. Finally, when q is incremented by s along the **then**-branch of the conditional, the new value of qq should be $d \cdot q = d \cdot (q' + s)$, where q' denotes the old value of q. Fortunately, the new value $d \cdot (q' + s)$ is

---

[8]See, for example, [Allen69] and [CockeKennedy77], where this process is termed "operator strength reduction."

[9]See Section 2.4.

equal to the current value of $qq+ss$ and does not require multiplication to compute. The program, extended with the new relations, is

```
assert qq=d·q, ss=d·s in
P₁⁺ : begin comment improved real–division program
   B₁⁺ : assert 0≤c<d, e>0
   (q,qq,s,ss) := (0,0,1,d)
   loop L₁⁺ : assert d·q≤c, c<d·(q+s)
          until s≤e
          (s,ss) := (s/2,ss/2)
          if qq+ss≤c then (q,qq) := (q+s,qq+ss) fi
          repeat
   E₁⁺ : assert q≤c/d, c/d<q+e
end
```

5.3. Examples

In this section, we apply the rules described above to the synthesis of several programs. We begin with the construction of a program to find the smallest element of an array. (A trace of the automatic synthesis of this program appears in Appendix 5.) The second example is the computation of the integer square-root function. In the third example we use "backward" synthesis rules to construct a program schema. Our concluding example is the *Partition* algorithm[10] which requires a greater degree of understanding and ingenuity to program. In these examples, we concentrate on showing a successful way of synthesizing the desired program; an implemented system would, of course, encounter many failures along the route and would need to retrace its steps and try alternatives.

---

[10] [Hoare61].

5.3.1. Array Minimum

In this example, we synthesize a program to search for the position of a minimal element in an array segment. The specification is

P_5: **begin comment** *minimum–position specification*
    assert $i \in \mathbf{Z}$, $n \in \mathbf{Z}$, $i \leq n$
    achieve $A[k] \leq A[i:n]$, $k \in [i,n]$ **varying** k
    end

The output specification consists of two parts: the value of the output variable k should be an integer within the bounds $[i:n]$ of the input array A, i.e. $1 \leq k \leq n$, and the value of the kth array element $A[k]$ should be less than or equal to all the (other) elements in $A[i:n]$. The conjunct $A[k] \leq A[i:n]$ is short for $(\forall \xi \in [i:n]) A[k] \leq A[\xi]$; in general, for a predicate α, we use $\alpha[u:v]$ as an abbreviation for $(\forall \xi \in [u:v]) \alpha(\xi)$. Note that the array A is a constant and only the value of the variable k may be altered by the program. The input specification asserts that the values of the input variables i and n will be integers (i.e. will be in the set $\mathbf{Z}$), with the added restriction that $i \leq n$. Thus, the synthesized program is to set the variable k to the index of an occurrence of the smallest element within the nonempty array segment $A[i:n]$.

5.3.1.1. Goal Splitting

The first step in synthesizing many iterative programs is to introduce new variables. Introducing a new variable will allow the program to manipulate its value so that the original goal may be achieved in stages. Using the

fact $p(\overline{u})$ **when** $p(\overline{v})$, $\overline{u} = \overline{v}$,

our goal may be strengthened (using the *strengthening rule* $<$s$>$) by introducing a new program variable j and substituting it for the constant i; the conjunct $j = i$ must be added to the goal. That gives us the new subgoal

> **purpose** $A[k] \leq A[i:n]$, $k \in [i:n]$
>   **achieve** $A[k] \leq A[j:n]$, $k \in [j:n]$, $j = i$ **varying** k,j
> **assert** $A[k] \leq A[i:n]$, $k \in [i:n]$.

The next step is to break the conjunctive goal into two parts: the first part will take advantage of the flexibility afforded by the new variable to achieve the relations for some values; after that, the more difficult task of achieving the goal for the input values is undertaken. Using the *protection rule* $<$p$>$, we first attempt to achieve $A[k] \leq A[j:n]$ with $k \in [j:n]$, and then to achieve $j = i$:

> **achieve** $A[k] \leq A[j:n]$, $k \in [j:n]$ **varying** k,j
> **achieve** $j = i$ **protecting** $A[k] \leq A[j:n]$, $k \in [j:n]$ **varying** k,j.

The first subgoal

> **achieve** $A[k] \leq A[j:n]$, $k \in [j:n]$ **varying** k,j

can be matched with the

> **fact** $p[u:u]$ **when** $p(u)$,

if we insist that $j = n$. So this subgoal may be replaced (using $<$s$>$) by the simpler

> **achieve** $A[k] \leq A[n]$, $j = n$, $k \in [j:n]$ **varying** k,j.

Using reflexivity,

> **fact** $u \leq u$,

to achieve "less-than-or-equal" by achieving "equal," we get

> **achieve** $A[k] = A[n]$, $j = n$, $k \in [j:n]$ **varying** k,j.

Now the fact that

> **fact** $f(u) = f(v)$ **when** $u = v$

for any function f (the array A may be construed as a function), suggests replacing $A[k] = A[n]$ by the sufficient $k = n$, leaving as our goal

> **achieve** $k = n$, $j = n$ **varying** k,j.

The latter goal is achievable (using the *assignment rule* $<$a$>$) by the multiple assignment

$$(k,j) := (n,n).$$

This, then, will be the first statement of the program.

5.3.1.2. Loop Formation

At this point, we are left with the second subgoal

achieve $j=i$ **protecting** $A[k] \leq A[j:n]$, $k \in [j:n]$ **varying** k,j.

The *forward-loop rule* $<f>$, suggests turning the protected relations into loop invariants, while repeating the loop until the goal is attained. The result of applying the rule is

> **loop**   **assert** $A[k] \leq A[j:n]$, $k \in [j:n]$
> **until** $j=i$
> **approach** $j=i$ **protecting** $A[k] \leq A[j:n]$, $k \in [j:n]$
>   **varying** k,j
> **repeat**
> **assert** $A[k] \leq A[j:n]$, $k \in [j:n]$, $j=i$.

The loop invariants $A[k] \leq A[j:n]$ and $k \in [j:n]$ are initialized by the assignment statement $(k,j) := (n,n)$ that precedes the loop. The loop-body subgoal is

> **approach** $j=i$ **protecting** $A[k] \leq A[j:n]$, $k \in [j:n]$
>   **varying** k,j,

meaning that one iteration of the loop should make some progress towards satisfying the exit test, while protecting the validity of the loop invariants. By noting that upon entering the loop the value of j is n and upon exiting the loop its value is i, where it is given that $i \leq n$, the *termination rule* $<t>$ suggests that the variable j remain an integer and decrease *monotonically* from n to i. The range of j is therefore $[i:n]$, and we have as our new loop-body goal:

> **assert** $j \neq i$
> **achieve** $j < j'$ **protecting** $A[k] \leq A[j:n]$, $k \in [j:n]$, $j \in [i:n]$
>   **varying** k,j,

where j' denotes the value of the variable j just prior to this goal. In other words, assuming that the goal $j = i$ has not yet been achieved, we must decrease j while protecting the loop invariants $A[k] \leq A[j:n]$ and $k \in [j:n]$. The additional requirement $j \in [i:n]$ is designed to guarantee that the loop will eventually terminate, since j cannot be decremented past i.

We are assuming that the protected invariants hold for the old value j' of j. In particular $j' \in [i:n]$, that is, $j' \in \mathbf{Z}$, $j' \geq i$, and $j' \leq n$. At the same time we know that for the loop to be continued it must be that $j' \neq i$, from which it follows that $j' - 1 \geq i$. So, in order to achieve $j \geq i$ as part of $j \in [i:n]$, it suffices to achieve $j \geq j' - 1$. This, together with the additional requirements that $j < j'$ and $j \in \mathbf{Z}$, forces j to be $j' - 1$. Our current goal, then, may be replaced by

achieve $j = j' - 1$ **protecting** $A[k] \leq A[j:n]$, $k \in [j:n]$, $j \in [i:n]$
varying k, j,

which is the same as saying

assert $A[k] \leq A[j:n]$, $k \in [j:n]$, $j \in [i:n]$
achieve $j = j' - 1$, $A[k] \leq A[j:n]$, $k \in [j:n]$, $j \in [i:n]$ **varying** k, j.

We have broken the protected clause into an assertion that the conjuncts held for the previous values of j and k and a goal to reachieve them for new values. Applying the *disjoint goal rule* <d> to separate out the (easily achievable) first conjunct $j = j' - 1$, followed by the *assignment rule* <a>, gives

assert $A[k] \leq A[j:n]$, $k \in [j:n]$, $j \in [i:n]$
$j := j - 1$
assert $A[k] \leq A[j + 1:n]$, $k \in [j + 1:n]$, $j + 1 \in [i:n]$
achieve $A[k] \leq A[j:n]$, $k \in [j:n]$ **varying** k.

The second assertion contains the information of the first assertion, taking into account the effect of the assignment statement.[11]

---

[11]See Chapter 6 for details of assertion propagation.

5.3.1.3. Conditional Formation

Part of the goal $A[k] \leq A[j:n]$ already holds (i.e. $A[k] \leq A[j+1:n]$), and part remains to be achieved (i.e. $A[k] \leq A[j]$). Using the following basic fact about universal quantification:

fact $p[u:v]$ **when** $p[u:w]$, $p[w+1:v]$, $w \in [u:v]$,

we can separate the goal into those two parts (this again is the synthesis rule $<s>$):

purpose $A[k] \leq A[j:n]$, $k \in [j:n]$
    achieve $A[k] \leq A[j:w]$, $A[k] \leq A[w+1:n]$, $w \in [j:n]$, $k \in [j:n]$
        varying k,w
assert $A[k] \leq A[j:n]$, $k \in [j:n]$

Since we have already asserted $A[k'] \leq A[j+1:n]$, we may achieve $A[k] \leq A[w+1:n]$ by leaving k unchanged (i.e. $k'=k$) and letting $w=j$. This also satisfies the requirements that $w \in [j:n]$ and $k \in [j:n]$, so we are left with only

achieve $A[k] \leq A[j]$ **varying** .

This last goal has an *empty* variable list; it can only be achieved by proving it true or testing that it is. It cannot be proved true, so we use the *conditional rule* $<c>$ to generate the program fragment

if $A[k] \leq A[j]$ **then**
    else   **assert** $A[k] \leq A[j+1:n]$, $k \in [j+1:n]$,
                    $A[k] > A[j]$
        achieve $A[k] \leq A[j:w]$, $A[k] \leq A[w+1:n]$,
                    $w \in [j:n]$, $k \in [j:n]$
            varying k,w **fi.**

The **then**-clause is empty, since $A[k] \leq A[j]$ holds at that point by virtue of the test; when the conditional test is false, we may use that fact, along with the fact that the invariants held for the prior value of j in an alternative attempt to achieve the previous goal. We know, then, that $A[j] < A[k'] \leq A[j+1:n]$, so to achieve $A[k] \leq A[w+1:n]$ we let $k=j=w$. Substituting for k and w, we have

> **achieve** $k=j=w$, $A[j] \leq A[j:j]$, $A[j] \leq A[j+1:n]$, $j \in [j:n]$
> **varying** k,w.

Since $A[k] \leq A[j+1:n]$ and $j+1 \leq n$ have already been asserted, this reduces to just

> **achieve** $k=j$ **varying** k.

Achieving this via an assignment statement, we obtain the conditional

> **if** $A[k] \leq A[j]$ **then else** $k := j$ **fi**,

or simply

> **if** $A[j] < A[k]$ **then** $k := j$ **fi**.

5.3.1.4. Extension

   All told, we have derived the following program for finding a minimum:

```
P₅: begin comment minimum–position program
  B₅: assert i ∈ Z, n ∈ Z, i ≤ n
  (k,j) := (n,n)
  loop L₅: assert A[k] ≤ A[j:n], k ∈ [j:n], j ∈ [i,n]
    until j=i
    j := j -1
    if A[j] < A[k] then k := j fi
    repeat
  E₅: assert A[k] ≤ A[i:n], k ∈ [i:n]
  end
```

Now suppose that we really want to

> **achieve** $A[k] \leq A[i:n]$, $k \in [i:n]$, $m=A[k]$ **varying** k,m.

The synthesized program only achieves the first two of these conjuncts; we need therefore to extend it to achieve the additional conjunct $m=A[k]$. One simple way to accomplish this would be to split the new goal into the two disjoint goals

achieve $A[k] \leq A[i:n]$, $k \in [i:n]$ **varying** k
achieve $m = A[k]$ **varying** m

(using rule $<\!d\!>$). For the first, we already have a program; for the second, we may simply append the assignment

$$m := A[k].$$

A second possibility would be to begin by achieving the relation $m = A[k]$, and then protecting that relation while achieving $A[k] \leq A[i:n]$ via P_5. In other words the relation $m = A[k]$ should be a global invariant of P_5, holding throughout execution of the program. In order to accomplish this goal, viz.

achieve $m = A[k]$ **in** P_5 **varying** m,

we must set m to the appropriate value whenever the variable k changes value. The assignments to k in the program are

$$k := n \mid k := j.$$

When k is initialized to n, we must initialize $m = A[k]$ to $A[n]$; when k is reset to j, we reset m to $A[j]$. This yields the extended program

$$(k,j,m) := (n,n,A[n])$$
loop L_5: **assert** $A[k] \leq A[j:n]$, $k \in [j:n]$
    until $j = i$
    $j := j - 1$
    if $A[j] < A[k]$ **then** $(k,m) := (j,A[j])$ **fi**
    repeat.

As it stands now, the first alternative requires less computation. But since we have established the global invariant $m = A[k]$, the conditional test $A[j] < A[k]$ may be simplified to $A[j] < m$. The final version of this program would then be

---

assert $m = A[k]$ **in**
P_5^+ : **begin comment** *minimum program*
  B_5^+ : **assert** $i \in \mathbf{Z}$, $n \in \mathbf{Z}$, $i \le n$
  $(k,j,m) := (n,n,A[n])$
  loop L_5^+ : **assert** $A[k] \le A[j:n]$, $k \in [j:n]$, $j \in [i,n]$
      until $j = i$
      $j := j - 1$
      if $A[j] < m$ **then** $(k,m) := (j,A[j])$ **fi**
      repeat
  E_5^+ : **assert** $A[k] \le A[i:n]$, $k \in [i:n]$, $m = A[k]$
  end

In a similar manner, we could begin with the program

---

assert $m = A[k]$ **in**
P_5' : **begin comment** *minimum–value program*
  B_5' : **assert** $i \in \mathbf{Z}$, $n \in \mathbf{Z}$, $i \le n$
  $(j,m) := (n,A[n])$
  loop L_5' : **assert** $m \le A[j:n]$, $j \in [i,n]$
      until $j = i$
      $j := j - 1$
      if $A[j] < m$ **then** $m := A[j]$ **fi**
      repeat
  E_5' : **assert** $m \le A[i:n]$
  end

---

that only achieves $m \le A[i:n]$, and extend it to achieve $m = A[k]$ as well. There is no easy way to set k at the end of P_5' so that $m = A[k]$. But we can globally

$$\textbf{achieve } m = A[k] \textbf{ in } P_5' \textbf{ varying } k$$

by examining the assignments to m,

$$m := A[n] \mid m := A[j].$$

The corresponding assignments to k would be

$$k := n \mid k := j,$$

yielding the same program P_2^+ .

5.3.2. Integer Square-Root

In Section 4.2.3 we instantiated a linear-search schema to compute integer square-roots, and in Section 2.7 we developed a binary integer square-root program from a binary-search schema. In this section, our goal is to synthesize a program satisfying the same specification,

P_4: **begin comment** *integer square-root specification*
    assert $a \in \mathbf{N}$
    achieve $z = \lfloor \sqrt{a} \rfloor$ **varying** z
    end

from scratch. The program should set the variable z to the largest integer not greater than the square-root of a, for any nonnegative integer a.

5.3.2.1. Goal Splitting

We assume that the $\sqrt{}$ function is not primitive; otherwise we could achieve our goal using the *assignment rule* $<$a$>$ to obtain

$$z := \lfloor \sqrt{a} \rfloor.$$

Therefore, as a first step, we endeavor to replace the goal with one that does not contain the $\sqrt{}$ function. Using the definition of $\lfloor u \rfloor$,

    fact $v = \lfloor u \rfloor$ **is** $v \leq u$, $u < v+1$, $v \in \mathbf{Z}$,

the goal

    achieve $z = \lfloor \sqrt{a} \rfloor$ **varying** z

`may be transformed (via rule $<$s$>$) into the equivalent goal

    purpose $z = \lfloor \sqrt{a} \rfloor$
        achieve $z \leq \sqrt{a}$, $\sqrt{a} < z+1$, $z \in \mathbf{Z}$ **varying** z
    assert $z = \lfloor \sqrt{a} \rfloor$.

Now to eliminate the $\sqrt{}$ operator, we use the

    fact $u \leq \sqrt{v}$ **is** $u^2 \leq v$ **when** $u \geq 0$, $v \geq 0$.

Accordingly, the conjunct $z \leq \sqrt{a}$ may be replaced by $z^2 \leq a$ and

$\sqrt{a} < z+1$ may be replaced by $a < (z+1)^2$, with the side conditions $z \geq 0$ and $z+1 \geq 0$ added (it is given that $a \geq 0$):

> **achieve** $z^2 \leq a$, $z \geq 0$, $a < (z+1)^2$, $z+1 \geq 0$, $z \in \mathbf{Z}$ **varying** z.

This simplifies to just

> **achieve** $z^2 \leq a$, $a < (z+1)^2$, $z \in \mathbf{N}$ **varying** z.

The above subgoal is a conjunction of three relations; the *protection rule* $<$p$>$ suggests splitting it into the two consecutive subgoals

> **purpose** $z^2 \leq a$, $a < (z+1)^2$, $z \in \mathbf{N}$ **varying** z
> **achieve** $a < (z+1)^2$, $z \in \mathbf{N}$ **varying** z
> **achieve** $z^2 \leq a$ **protecting** $a < (z+1)^2$, $z \in \mathbf{N}$ **varying** z
> **assert** $z^2 \leq a$, $a < (z+1)^2$, $z \in \mathbf{N}$ **varying** z.

Later, we shall see what alternative splittings might result in.

To achieve the first subgoal

> **achieve** $a < (z+1)^2$, $z \in \mathbf{N}$ **varying** z

we apply the *strengthening rule* $<$s$>$ to introduce an additional variable, using the transitivity of inequality expressed in the

> **fact** $u < w$ **when** $u < v$, $v \leq w$.

The result is

> **achieve** $a < v$, $v \leq (z+1)^2$, $z \in \mathbf{N}$ **varying** z,v.

Now, the

> **fact** $u \leq u^2$ **when** $u \geq 1 \lor u \leq 0$

tells us that taking $z+1$ for v will give $v \leq (z+1)^2$, provided that $z+1 \geq 1 \lor z+1 \leq 0$. The subgoal $z \in \mathbf{N}$ implies $z+1 \geq 1$, leaving

> **achieve** $v = z+1$, $a < v$, $z \in \mathbf{N}$ **varying** z,v.

Since v is not an output variable, it may be eliminated from the goal, substituting $z+1$ for it. The goal

> **achieve** $a < z+1$, $z \in \mathbf{N}$ **varying** z

that remains can be strengthened further to

achieve $a = z$ **varying** z,

by matching it with the

fact $u < u + v$ **when** $v > 0$.

The goal, in turn, may be attained by the simple assignment

purpose $a < (z+1)^2$, $z \in \mathbf{N}$
$z := a$
assert $z = a$.

5.3.2.2. Loop Formation

The *forward loop rule* $<f>$ suggests turning the second subgoal

achieve $z^2 \leq a$ **protecting** $a < (z+1)^2$, $z \in \mathbf{N}$ **varying** z

into a loop with the invariant

assert $a < (z+1)^2$, $z \in \mathbf{N}$

maintained true until the exit clause

until $z^2 \leq a$

becomes true. We have the following skeleton of a loop:

purpose $a < (z+1)^2$, $z \in \mathbf{N}$
$z := a$
assert $z = a$
purpose $z^2 \leq a$, $a < (z+1)^2$, $z \in \mathbf{N}$
  loop L_4': **assert** $a < (z+1)^2$, $z \in \mathbf{N}$
    until $z^2 \leq a$
    approach $z^2 \leq a$ **protecting** $a < (z+1)^2$, $z \in \mathbf{N}$
             varying z
  repeat
E_4': **assert** $z^2 \leq a$, $a < (z+1)^2$, $z \in \mathbf{N}$.

Within the loop body, we must

approach $z^2 \leq a$ **protecting** $a < (z+1)^2$, $z \in \mathbf{N}$ **varying** z

in order to make progress towards the exit test, while protecting the two invariants.

To ensure termination of this loop, the *termination rule* $<t>$ requires that the nonnegative integer z be reduced in some well-founded ordering. We note that upon exit $z^2 \leq a$, while upon entering the loop $z = a$. It follows that the entry value of z is greater than its exit value, and we hypothesize that z decreases monotonically from a to it final value. Taking the set of natural numbers $\mathbf{N}$, under their natural ordering $>$, as the well-founded set, we obtain the loop-body subgoal

> **assert** $a < z^2$
> **achieve** $z < z'$ **protecting** $a < (z+1)^2$, $z \in \mathbf{N}$ **varying** z.

That is, we wish to set the nonnegative integer z to a value less than its current one, while protecting the loop invariant $a < (z+1)^2$. The assertion indicates that the exit test $z^2 \leq a$ does not yet hold.

With each loop iteration we wish to decrease the value of z, while protecting the invariant $a < (z+1)^2$. Using the transitivity of inequality again suggests looking for some v such that $a < v$ and $v \leq (z+1)^2$. In particular, $a < z'^{\,2}$ is true for the previous value of z; therefore, to achieve $a < (z+1)^2$, we need only achieve $z'^{\,2} \leq (z+1)^2$, i.e. $z' \leq z+1$. This leaves us with the goal

> **achieve** $z < z'$, $z' \leq z+1$ **protecting** $z \in \mathbf{N}$ **varying** z

which is equivalent to

> **achieve** $z = z' - 1$ **varying** z

and may be achieved by the assignment

$$z := z - 1.$$

We have derived the following program, with most of the **purpose** statements retained:

P_4': **begin comment** *cluttered integer square–root program*
  B_4': **assert** $a \in \mathbf{N}$
  purpose $z = \lfloor \sqrt{a} \rfloor$
    purpose $z \leq \sqrt{a}$, $\sqrt{a} < z + 1$, $z \in \mathbf{N}$
      purpose $z^2 \leq a$, $a < (z+1)^2$, $z \in \mathbf{N}$
        purpose $a < (z+1)^2$, $z \in \mathbf{N}$
          $z := a$
        assert $z = a$
        purpose $z^2 \leq a$, $a < (z+1)^2$, $z \in \mathbf{N}$
          loop L_4': **assert** $a < (z+1)^2$, $z \in \mathbf{N}$
              until $z^2 < a$
              purpose $z < z'$, $a < (z+1)^2$, $z \in \mathbf{N}$
                purpose $z < z'$, $z' \leq z + 1$, $z \in \mathbf{N}$
                  $z := z - 1$
                assert $z = z' - 1$, $z \in \mathbf{N}$
              assert $z < z'$, $a < (z+1)^2$, $z \in \mathbf{N}$
              repeat
          assert $z^2 \leq a$, $a < (z+1)^2$, $z \in \mathbf{N}$
        assert $z^2 \leq a$, $a < (z+1)^2$, $z \in \mathbf{N}$
      assert $z \leq \sqrt{a}$, $\sqrt{a} < z + 1$, $z \in \mathbf{N}$
  E_4': **assert** $z = \lfloor \sqrt{a} \rfloor$
  end

5.3.2.3. Alternatives

What would have happened had we split the goal

      achieve $z^2 \leq a$, $a < (z+1)^2$, $z \in \mathbf{N}$ **varying** z

in a different manner? Had we split it into

    achieve $z \in \mathbf{N}$ **varying** z
    achieve $z^2 \leq a$, $a < (z+1)^2$ **protecting** $z \in \mathbf{N}$ **varying** z,

we would have been led to the subgoals

> **achieve** $z \in N$ **varying** z
> **loop** L_4'': **assert** $z \in N$
>   **until** $z^2 \leq a \wedge a < (z+1)^2$
>   **approach** $z^2 \leq a$, $a < (z+1)^2$ **protecting** $z \in N$
>   **varying** z
> **repeat**.

The choice of initial value for z satisfying $z \in N$ is completely arbitrary; but to ensure termination, we would have to determine in which direction to change z within the loop. The resulting program might look like

> P_4'': **begin comment** *nondeterministic integer square-root*
> B_4'': **assert** $a \in N$
> $z :\in N$
> **loop** L_4'': **assert** $z \in N$
>   **until** $z^2 \leq a \wedge a < (z+1)^2$
>   **if** $z^2 \leq a$ **then** $z := z+1$ **else** $z := z-1$ **fi**
>   **repeat**
> E_4'': **assert** $z = \lfloor \sqrt{a} \rfloor$
> **end**

where $z :\in N$ is a nondeterministic assignment of some element of the set N to the variable z. This solution is more complicated than the previous one; in general, it is advisable to maintain invariant as much of the goal as possible and keep the exit test as simple as possible.

Had we chosen, instead, to first

$$\textbf{achieve } z^2 \leq a, \; z \in N \textbf{ varying } z$$

and then

$$\textbf{achieve } a < (z+1)^2 \quad \textbf{protecting } z^2 \leq a, \; z \in N$$
$$\textbf{varying } z,$$

we would be led, in a manner similar to our first try, to the loop skeleton

achieve $z^2 \leq a$, $z \in \mathbf{N}$ **varying** z
purpose $z^2 \leq a$, $a < (z+1)^2$, $z \in \mathbf{N}$
  loop L_4: **assert** $z^2 \leq a$, $z \in \mathbf{N}$
    until $a < (z+1)^2$
    approach $a < (z+1)^2$ **protecting** $z^2 \leq a$, $z \in \mathbf{N}$
                varying z
  repeat
assert $z^2 \leq a$, $a < (z+1)^2$, $z \in \mathbf{N}$.

To achieve the initialization subgoal it suffices to

$$\textbf{achieve } z^2 \leq 0, \ z \in \mathbf{N} \textbf{ varying } z,$$

since a is nonnegative. Given the

$$\textbf{fact } 0 \leq u^2,$$

it must be that $z^2 = 0$, so we may initialize $z := 0$. We would then decide to approach the exit test by *increasing* z from 0. To guarantee termination, we could use the well-founded set of integers less than $\sqrt{a}$; the smaller the integer, the greater it is in the well-founded ordering. Continuing in a manner paralleling the derivation of $P_4{}'$, we get

P_4: **begin comment** *simple integer square-root program*
  B_4: **assert** $a \in \mathbf{N}$
  $z := 0$
  loop L_4: **assert** $z^2 \leq a$, $z \in \mathbf{N}$
      until $a < (z+1)^2$
      $z := z+1$
      repeat
  E_4: **assert** $z = \lfloor \sqrt{a} \rfloor$
  end

5.3.2.4. Extension

In the above program P_4, the exit test $a < (z+1)^2$ is relatively difficult to compute, as it involves squaring. It may be replaced by the simpler $a < u$, however, if we extend the program to achieve the global invariant $u = (z+1)^2$:

achieve $u=(z+1)^2$ in P_4 varying u.

The variable z is set by two assignments in the program; whenever they are executed, we must assign appropriate values to u to maintain the desired relation $u=(z+1)^2$.

One way to accomplish this is to use rules that relate assignment statements to the values of program variables. Such rules are also used for program annotation in Chapter 6. One of them (number $<15>$ in Appendix 4) states that a global assertion of the form

assert $(y-b_0)\cdot 2\cdot a_2^2 = (x-a_0)\cdot[b_1\cdot(x-a_0-a_2)+2\cdot a_2\cdot(b_2+b_1\cdot a_0)]$ in P

holds if all the assignments to x and y in P are of one of the two forms

$$(x,y) := (a_0,b_0) \mid (x,y) := (x+a_2,y+b_1\cdot x+b_2),$$

where a_0, b_0, a_2, b_1, and b_2 are of constant value within P.

To apply this rule to the problem at hand, we match x and y in the rule with z and u in the program, respectively. The assignments to z are

$$z := 0 \mid z := z+1,$$

so we must let $a_0 \Rightarrow 0$ and $a_2 \Rightarrow 1$. Thus, assignments of the form

$$(z,u) := (0,b_0) \mid (z,u) := (z+1,u+b_1\cdot z+b_2)$$

achieve the relation

$$(u-b_0)\cdot 2\cdot 1^2 = (z-0)\cdot[b_1\cdot(z-0-1)+2\cdot 1\cdot(b_2+b_1\cdot 0)],$$

i.e.

$$(u-b_0)\cdot 2 = z\cdot[b_1\cdot(z-1)+2\cdot b_2].$$

We look, therefore, for instantiations of b_0, b_1, and b_2, such that

$$(u-b_0)\cdot 2 = z\cdot[b_1\cdot(z-1)+2\cdot b_2] \quad \Rightarrow \quad u=(z+1)^2.$$

Isolating u to the left of the equality and matching, leaves

$$z\cdot[b_1\cdot(z-1)/2+b_2]+b_0 \quad \Rightarrow \quad (z+1)^2.$$

Transforming $(z+1)^2$ into $z^2+2\cdot z+1 = z\cdot(z+2)+1$, suggests $b_0 \Rightarrow 1$ and $b_1\cdot(z-1)/2+b_2 \Rightarrow z+2$, which, in turn, suggests $b_1 \Rightarrow 2$ and

$b_2 \Rightarrow 3$. Instantiating b_0, b_1, and b_2 back into the assignments, we get

$$(z,u) := (0,1) \mid (z,u) := (z+1, u+2 \cdot z + 3).$$

A further improvement would be to

achieve $v = 2 \cdot z + 3$ **in** P_4.

By another rule ($<11>$ in Appendix 4), to get

assert $a_1 \cdot (y - b_0) = b_1 \cdot (x - a_0)$ **in** P,

the proper assignments are

$$(x,y) := (a_0, b_0) \mid (x,y) := (x + a_1 \cdot u, y + b_1 \cdot u).$$

This suggests the instantiation

$$\begin{array}{ccc} x & \Rightarrow & z \\ y & \Rightarrow & v \\ a_0 & \Rightarrow & 0 \\ a_1 & \Rightarrow & 1 \\ u & \Rightarrow & 1, \end{array}$$

leaving

$$(v - b_0) = b_1 \cdot z \quad \Rightarrow \quad v = 2 \cdot z + 3.$$

Taking $b_0 \Rightarrow 3$ and $b_1 \Rightarrow 2$, we get the assignments

$$(z,u,v) := (0,1,3) \mid (z,u,v) := (z+1, u+v, v+2)$$

and the program

```
assert u=(z+1)², v=2·z+3, z∈N in
P₄⁺ : begin comment integer square-root program
      (z,u,v) := (0,1,3)
      loop   L₄⁺ : assert z² ≤ a
             until a < u
             (z,u,v) := (z+1, u+v, v+2)
             repeat
      E₄⁺ : assert z=⌊√a⌋
      end
```

5.3.2.5. Binary Integer Square-Root

In Section 2.7, we outlined the synthesis of the following binary integer square-root program, based on a binary-search schema:

```
assert a ∈ N, z ∈ N, s ∈ 2^N in
P₂: begin comment binary integer square–root program
   B₂: assert a ∈ N
   z := 0
   purpose a < s², s ∈ 2^N
      s := 1
      loop    L₂': assert s ∈ 2^N
              until a < s²
              s := 2·s
              repeat
   purpose z² ≤ a, a < (z+s)², s ≤ 1
      loop    L₂: assert z² ≤ a, a < (z+s)²
              until s ≤ 1
              s := s/2
              if (z+s)² ≤ a then z := z+s fi
              repeat
   E₂: assert z=⌊√a⌋
   end
```

In this program, the exit test $a < s^2$ and conditional test $(z+s)^2 \le a$ are the most expensive expressions computed. By achieving the goal

$$\text{achieve } w = s^2 \text{ in } P_2 \text{ varying } w,$$

we can replace the test $a < s^2$ with $a < w$. Whenever s is updated, the new variable w must be updated correspondingly so that the relation with s remains invariant. When s is doubled, w must be quadrupled; when s is halved, w must be quartered.

Since the expression $(z+s)^2$ is equivalent to $z^2 + 2 \cdot s \cdot z + s^2 = z^2 + 2 \cdot s \cdot z + w$, the conditional test can be simplified to $u + v + w$ by achieving the additional global invariants

$$\text{achieve } u = z^2, \ v = 2 \cdot s \cdot z \text{ in } P_2 \text{ varying } u, v.$$

Updating u and v whenever z or s is assigned to, results in the program

$$(z,s,u,v,w) := (0,1,0,0,1)$$
loop L_2': **assert** $s \in 2^N$
    until $a < w$
    $(s,w) := (2{\cdot}s, 4{\cdot}w)$
    repeat
loop L_2: **assert** $z^2 \le a$, $a < (z+s)^2$
    until $s \le 1$
    $(s,v,w) := (s/2, v/2, w/4)$
    if $u+v+w \le a$
        then $(z,u,v) := (z+s, u+v+w, v+2{\cdot}w)$ **fi**
repeat.

Since the variable z affects only the value of z itself, it is a logical candidate for elimination from the program. The only problem is that z is to contain the final result. Fortunately, if $v = 2{\cdot}s{\cdot}z$ holds throughout, then the value of z may be recovered from the values of v and s. Since $s \in 2^N$ globally, when $s \le 1$ upon exit, it must be that s is equal 1. Thus, the goal

achieve $v = 2{\cdot}s{\cdot}z$ **varying** z

may be achieved by assigning the value $v/(2{\cdot}s) = v/2$ to z at the end of the program.

Once the assignment $z := z+s$ has been eliminated, the variable s only affects the exit test $s \le 1$. But we can replace the variable s with $\sqrt{w}$, since $w = s^2$ and w and s are known to be positive. This gives us the exit clause

until $\sqrt{w} \le 1$,

for which the primitive

until $w \le 1$

may be substituted.

Squeezing the last drop of ink out of this example, we can make yet another slight improvement: to simplify the test $u+v+w \le a$, we can apply the transformation $u \Rightarrow u+a$, yielding the conditional

if $u+v+w \le 0$ **then** $(u,v) := (u+v+w, v+2{\cdot}w)$ **fi**

and initialization

$$(u,v,w) := (-a,0,1).$$

To avoid recomputing the expression $u+v+w$ for both the conditional test and **then**-branch assignment, a temporary variable could be generated, say t, such that $t = u+v+w$, which would then be used to test $t \leq 0$ and assign $(u,v) := (t,v+2 \cdot w)$. Since $u = z^2$ and $v = 2 \cdot s \cdot z$ do not change value within the first loop (when $z = 0$), they can just as well be set upon entering the second loop. Incorporating these improvements, we derive the program

$$w := 1$$
$$\textbf{loop } L_2': \textbf{ assert } \sqrt{w} \in 2^N$$
$$\qquad \textbf{until } a < w$$
$$\qquad w := 4 \cdot w$$
$$\qquad \textbf{repeat}$$
$$(u,v) := (-a,0)$$
$$\textbf{loop } L_2: \textbf{ assert } v^2/(4 \cdot w) \leq a, \ a < v^2/(4 \cdot w) + v + w$$
$$\qquad \textbf{until } w \leq 1$$
$$\qquad (v,w) := (v/2, w/4)$$
$$\qquad t := u + v + w$$
$$\qquad \textbf{if } t \leq 0 \textbf{ then } (u,v) := (t, v+2 \cdot w) \textbf{ fi}$$
$$\qquad \textbf{repeat}$$
$$z := v/2.$$

The variables s and z have been replaced in the assertions by their equivalents $\sqrt{w}$ and $v/(2 \cdot \sqrt{w})$, respectively, and expressions have been simplified.

Finally, the two global transformations

$$w \ \Rightarrow \ w/2$$

and

$$v \ \Rightarrow \ v - w/2,$$

together, simplify the assignments $v := v + 2 \cdot w$ and $t := u + v + w$ to $v := v + w$ and $t := u + v$, respectively. In the process, the final assignment $z := v/2$ becomes $z := (v - w/2)/2$, but at that point $w = 2$, simplifying the right-hand side to $(v-1)/2$. The right-hand side of the assignment $v := v/2$ becomes $(v - w/2)/2 + w/8 = (v - w/4)/2$. So as not to compute $w/4$ twice, we may first assign $w := w/4$ and then

$v := (v - w)/2$. Accordingly, our final version of this program is

---

assert $a \in \mathbf{N}$, $w \in 2 \cdot 4^{\mathbf{N}}$, $2 \cdot w \cdot (u + a) = (v - w/2)^2$ **in**
P_2^+ : **begin comment** *optimized integer square-root program*
  B_2^+ : **assert** $a \in \mathbf{N}$
  $w := 2$
  loop   until $2 \cdot a < w$
        $w := 4 \cdot w$
        repeat
  $(u,v) := (-a, w/2)$
  loop L_2^+ : **assert** $u \le 0$, $0 < u + \sqrt{2 \cdot w \cdot (u + a)} + w/2$
        until $w \le 2$
        $w := w/4$
        $v := (v - w)/2$
        $t := u + v$
        if $t \le 0$ **then** $(u,v) := (t, v + w)$ **fi**
        repeat
  $z := (v - 1)/2$
  E_2^+ : **assert** $z = \lfloor \sqrt{a} \rfloor$
  end

We have successfully replaced the first exit test $a < s^2$ and conditional test $(z + s)^2 \le a$, both of which involve squaring (a relatively expensive operation), with the simpler tests $2 \cdot a < w$ and $t \le 0$. This, at the cost of updating variables by addition, subtraction, halving, quadrupling, and quartering (relatively cheap operations on binary computers).[12]

---

[12]Cf. the versions of the integer square-root program developed in [Dijkstra76] and [Blikle78]. [BroyKriegBruckner80] show how to develop an integer square-root program, starting from a recursive schema.

The real square-root program

```
Q₁: begin comment real square-root program
   B₁: assert a ≥ 0, e > 0
   (r,s) := (0, a+1)
   loop  L₁: assert r ≤ √a , √a < r+s
         until s ≤ e
         s := s/2
         if (r+s)² ≤ a then r := r+s fi
         repeat
   E₁: assert |√a - r| < e
   end
```

may be optimized in a similar manner. In this case, the initialization would need to be redone. The result is the following square-root algorithm:

```
assert 4·s·u = r² - a in
Q₁⁺: begin comment optimized real square-root program
   B₁⁺ : assert a ≥ 0, e > 0
   if a=1   then   r := 1
            else   (r,s,u) := (0, max(a/2,1/2), -min(a/2,1/2))
                   loop L₁⁺ : assert r ≤ √a , √a < r+2·s
                         until s ≤ e/2
                         (s,u) := (s/2, 2·u)
                         t := u + r + s
                         if t ≤ 0 then (r,u) := (r+2·s, t) fi
                         repeat                              fi
   E₁⁺ : assert |√a - r| < e
   end
```

which is essentially the same as that in [Wensley59].

5.3.3. Associative Recursion

In this example, we illustrate the use of the *preservation* and *backward loop* rules to synthesize a program schema for computing a recursive function that is defined in terms of another, associative, function. The goal is

S_7: **begin** *associative–recursion specification*
    assert $p(u) \supset f(u) = c(u),$
             $\sim p(u) \supset f(u) = h(f(e(u)), u),$
             $\sim p(u) \supset e(u) < u,$
             $h(u, h(v, w)) = h(h(u, v), w)$
    achieve $z = f(x)$ **varying** z
    end

where p is some predicate, c, h, and e are functions, and $e(x) < x$ in some well-founded ordering $>$. We assume that f is not primitive, while p, c, h, and e are.

5.3.3.1. Recursive Solution

Given that $f(x) = c(x)$ when $p(x)$, the above goal may be strengthened to

        purpose $z = f(x)$
            achieve $z = c(x)$, $p(x)$ **varying** z
        assert $z = f(x)$.

The *conditional rule* suggests solving this by testing for $p(x)$:

        purpose $z = f(x)$
          if $p(x)$ **then**   **assert** $p(x)$
                             achieve $z = c(x)$ **varying** z
                   else   **assert** $\sim p(x)$
                             achieve $z = f(x)$ **varying** z **fi**
        assert $z = f(x)$.

The **then**-goal may be achieved by the assignment

$$z := c(x),$$

since c is assumed to be primitive. The **else**-goal may be transformed into

$$\textbf{achieve } z = h(f(e(x)), x) \textbf{ varying } z,$$

since $\sim p(x)$ is assumed. This may be strengthened by adding a new program variable y for the nonprimitive expression $f(e(x))$:

$$\textbf{achieve } z{=}h(y,x), \; y{=}f(e(x)) \textbf{ varying } z,y,$$

which in turn may be split into

$$\textbf{achieve } y{=}f(e(x)) \textbf{ varying } y$$
$$\textbf{achieve } z{=}h(y,x) \textbf{ varying } z$$

At this point a recursive call can be inserted to achieve the first of the above disjoint goals, since $e(x) < x$ ensures termination of the recursive loop. The second disjoint goal may be achieved by an assignment. The resulting program is

---

$R_7(z,x)$: **procedure comment** *recursion schema*
  B_7: **assert** $p(u) \supset f(u){=}c(u),$
               $\sim p(u) \supset f(u){=}h(f(e(u)),u),$
               $\sim p(u) \supset e(u) < u,$
               $h(u,h(v,w)){=}h(h(u,v),w)$
  purpose $z{=}f(x)$
    if $p(x)$ **then**   **assert** $p(x)$
                      $z := c(x)$
           else   **assert** $\sim p(x), \; e(x) < x$
                     purpose $y{=}f(e(x))$
                     $R_7(y,e(x))$
                     assert $y{=}f(e(x))$
                     $z := h(y,x)$ **fi**
  E_7: **assert** $z{=}f(x)$
  end

5.3.3.2. Iterative Solution

The above recursive solution does not take advantage of the associativity of h. An alternative, more efficient, solution can be obtained by strengthening the goal

$$\textbf{achieve } z{=}h(f(e(x)),x) \textbf{ varying } z$$

of the **else**-branch in a different manner. This time we introduce two new variables y and s:

$\quad$ **achieve** $z = h(f(s), y)$, $y = x$, $s = e(x)$ **varying** z, y, s.

Using the *preservation rule* to split this goal, we get

$\quad$ **achieve** $y = x$, $s = e(x)$ **varying** y, s
$\quad$ **achieve** $z = h(f(s), y)$ **preserving** $z = h(f(s), y)$ **for** y, s
$\qquad$ **varying** z, y, s.

The first subgoal may be achieved by the assignment

$$(y, s) := (x, e(x)).$$

The second subgoal may be transformed (using the given properties of f) into

$\quad$ **achieve** $z = h(c(s), y)$, $p(s)$ **preserving** $z = h(f(s), y)$ **for** y, s
$\qquad$ **varying** z, y, s

and then split into the disjoint subgoals

$\quad$ **achieve** $p(s)$ **preserving** $z = h(f(s), y)$ **for** y, s
$\qquad$ **varying** y, s
$\quad$ **achieve** $z = h(c(s), y)$ **varying** z.

The second of these subgoals may be achieved by a straightforward assignment; the first subgoal is amenable to the *backward loop rule*, yielding the loop

$\quad$ **loop** L_7: **purpose** $z = h(f(s), y)$
$\qquad$ **until** $p(s)$
$\qquad$ **assert** $\sim p(s)$
$\qquad$ **approach** $p(s)$ **preserving** $z = h(f(s), y)$ **for** y, s
$\qquad\qquad$ **varying** y, s
$\quad$ **repeat**.

Applying the *backward termination rule* to the loop-body goal gives

$\quad$ **achieve** $s < s'$ **preserving** $z = h(f(s), y)$ **for** y, s
$\qquad$ **varying** y, s;

that is,

$\quad$ **achieve** $s < s'$, $z = h(f(s), y) \supset z = h(f(s'), y')$ **varying** y, s.

Since $\sim p(s')$, this is the same as

achieve $s < s'$, $z=h(f(s),y) \supset z=h(h(f(e(s'))),s'),y')$ **varying** y,s.

By taking note of the associativity of h, the two sides of the implication can be made to match. To

achieve $s < s'$, $z=h(f(s),y) \supset z=h(f(e(s')),h(s',y'))$ **varying** y,s,

we can

achieve $s < s'$, $y=h(s',y')$, $s=e(s')$ **varying** y,s

by assigning

$$(y,s) := (h(s,y),e(s)),$$

since $e(s) < s'$.

The schema we have derived is

S_7: **begin comment** *associative–recursion schema*
  B_7: **assert** $p(u) \supset f(u)=c(u)$,
        $\sim p(u) \supset f(u)=h(f(e(u)),u)$,
        $\sim p(u) \supset e(u) < u$,
        $h(u,h(v,w))=h(h(u,v),w)$
  if $p(x)$ **then**   **assert** $p(x)$
                $z := c(x)$
         else   **assert** $\sim p(x)$
                $(y,s) := (x,e(x))$
                loop L_7: **purpose** $z=h(f(s),y)$
                      until $p(s)$
                      $(y,s) := (h(s,y),e(s))$
                      repeat
                $z := h(c(s),y)$
                assert $z=h(c(s),y)$, $p(s)$ **fi**
  E_7: **assert** $z=f(x)$
  end

It applies, for example, to computing factorials, as specified by

---

P_6: **begin comment** *factorial specification*
  assert $n \in N$
  achieve $m=n!$ **varying** m
  end

To apply our schema S_7 to this problem, the appropriate instantiation is

$$
\begin{array}{rcl}
z & \Rightarrow & m \\
x & \Rightarrow & n \\
f(u) & \Rightarrow & u! \\
p(u) & \Rightarrow & u = 0 \\
c(u) & \Rightarrow & 1 \\
e(u) & \Rightarrow & u - 1 \\
h(u,v) & \Rightarrow & v \cdot u,
\end{array}
$$

and the instantiated program is

P_6: **begin comment** *iterative factorial program*
   B_6: **assert** $n \in \mathbf{N}$
  if $n = 0$ **then**   **assert** $n = 0$
                           $m := 1$
           else   **assert** $n \neq 0$
                         $(y,s) := (n, n-1)$
                         loop L_6: **purpose** $m = y \cdot s!$
                             until $s = 0$
                             $(y,s) := (y \cdot s, s-1)$
                             repeat
                         $m := y$
                         assert $m = y$, $s = 0$ **fi**
  E_6: **assert** $m = n!$
  end

The schema S_7 is similar to one of the recursion-elimination transformations in [DarlingtonBurstall76]. It can be specialized when the base-case value $c(x)$ is an identity element of h. In the factorial case, since $u \cdot 1 = 1 \cdot u = u$, the two branches of the conditional may be combined to yield a simpler program.[13]

---

[13]See also Sections 4.2.4 and 4.3.2.

5.3.4. Partition

In this last synthesis example, we consider the *Partition* problem: given an array segment $A[i:j]$, rearrange its elements so that some position g partitions the segment into two ordered parts. In other words, each element of the left part $A[i:g]$ is to be less than or equal to all elements of the right part $A[g+1:j]$. The goal may be expressed as

> P_8: **begin comment** *partition specification*
>   **assert** $i \in N$, $j \in N$, $i < j$
>   **achieve** $A[i:g] \leq A[g+1:j]$, $i \leq g$, $g+1 \leq j$,
> $\{A[i:j]\} = \{A'[i:j]\}$
>   **varying** A,g

where A' represents the input value of the array A and the notation $A[i:g] \leq A[g+1:j]$ means

$$(\forall u \in [i:g])(\forall v \in [g+1:j])\ A[u] \leq A[v].$$

The notation $\{A[u:v]\}$ denotes the multiset $\{A[u],A[u+1],\cdots,A[v]\}$; thus, the fourth conjunct of the goal implies that the new array segment is a permutation of the original segment.

5.3.4.1. First Solution

As a first try, we can strengthen the output specification, using the

fact $p[u:u]$ **when** $p(u)$

to eliminate the quantifier $[i:g]$. What we wish, then, is to

achieve $A[i] \leq A[g+1:j]$, $g=i$, $\{A[i:j]\}=\{A'[i:j]\}$ **varying** A,g.

The subgoals $i \leq g$ and $g+1 \leq j$ have been dropped since they follow from the new goal $g=i$ and the assertion $i < j$. The subgoal $g=i$ can be achieved by the assignment

$$g := i,$$

leaving only

achieve $A[i] \leq A[i+1{:}j]$, $\{A[i{:}j]\}=\{A'[i{:}j]\}$ **varying** A,

(i has replaced g in the remaining subgoal).

In Section 5.3.1 we synthesized a program to find the position of a minimal element of an array; so now we can use that program P_5 to

achieve $A[z] \leq A[i+1{:}j]$, $z \in [i+1{:}j]$ **varying** z.

To avail ourselves of the existing program, we first use the transitivity of inequality,

fact $u \leq v$ **when** $u \leq w$, $w \leq v$,

to transform our goal into

achieve $A[i] \leq w$, $w \leq A[i+1{:}j]$, $\{A[i{:}j]\}=\{A'[i{:}j]\}$ **varying** A,w.

We can fix the value of the new program variable w as we see fit, which in our case would be to $A[z]$. Splitting the resultant goal into the disjoint goals, we have

> **achieve** $A[z] \leq A[i+1{:}j]$, $z \in [i+1{:}j]$, $\{A[i{:}j]\}=\{A'[i{:}j]\}$
> **varying** z
> **achieve** $A[i] \leq A[z]$
> **protecting** $A[z] \leq A[i+1{:}j]$, $z \in [i+1{:}j]$,
> $\{A[i{:}j]\}=\{A'[i{:}j]\}$
> **varying** A.

The permutation requirement in the first goal is satisfied, since that goal does not vary A (i.e. $A'=A$); the rest is achieved by P_5. The remaining goal $A[i] \leq A[z]$ cannot be achieved by an assignment $A[z] := A[i]$, since that might violate the permutation requirement. We can, however, test if $A[i] \leq A[z]$ already holds:

> **if** $A[i] \leq A[z]$ **then**
> **else**   **assert** $A[i] > A[z]$
> **achieve** $A[i] \leq A[z]$ **protecting** $\cdots$
> **varying** A **fi**.

Knowing, for the **else**-branch, that $A'[z] < A'[i]$, suggests achieving $A[i] \leq A[z]$ by letting $A'[z]=A[i]$ and $A'[i]=A[z]$. Since $z \in [i{:}j]$, this also protects the permutation requirement $\{A[i{:}j]\}=\{A'[i{:}j]\}$. We have

      if $A[i] \leq A[z]$ **then**
                   else   **assert** $A[i] > A[z]$
                              $(A[i], A[z]) := (A[z], A[i])$ **fi**,

or, equivalently,

      if $A[i] > A[z]$ **then** $(A[i], A[z]) := (A[z], A[i])$ **fi**.

Accordingly, our first solution is

P_8': **begin comment** *first partition program*
  B_8': **assert** $i \in \mathbf{N}$, $j \in \mathbf{N}$, $i < j$
  $(z, j') := (j, j)$
  loop L_8': **assert** $A[z] \leq A[j':j]$, $z \in [j':j]$, $j' \in [i, j]$
      until $j' = i$
      $j' := j' - 1$
      if $A[j'] < A[z]$ **then** $z := j'$ **fi**
  repeat
  if $A[i] > A[z]$ **then** $(A[i], A[z]) := (A[z], A[i])$ **fi**
  E_8': **assert** $A[z] \leq A[i:j]$, $z \in [i:j]$
  end

This program leaves something to be desired. Despite the fact that it satisfies the stated specification and that it is not inefficient, the desire (for the usual applications of *Partition*) that g be closer to the mean of i and j went unspecified. The above program always results in g being equal to i.[14]

---

[14]This is a good example of the difficulty that can arise in attempting to fully specify the requirements for a program. Besides "correctness" (no element in the left half should be greater than any element in the right half), and "efficiency" (the number of comparisons should be on the same order as the number of elements), there is also a desideratum pertaining to the average result (that the mean position of the split be "near" the middle).

5.3.4.2. Second Solution

We do not want, then, to force $g = i$. Instead, we let g vary and reconsider our original goal (temporarily leaving out the permutation requirement)

achieve $A[i:g] \leq A[g+1:j]$, $i \leq g$, $g+1 \leq j$ **varying** A, g.

This time, we introduce a new variable h to replace the expression $g+1$:

achieve $A[i:g] \leq A[h:j]$, $g+1 = h$, $i \leq g$, $h \leq j$ **varying** A, g, h.

Then we try to eliminate the double (implicit) quantifier in $A[i:g] \leq A[h:j]$ by introducing another variable w, using transitivity as before:

achieve $A[i:g] \leq w$, $w \leq A[h:j]$, $g+1 = h$, $i \leq g$, $h \leq j$
    varying A, g, h, w.

There are now four variables for the program to set: A, g, h, and w.

The above conjunctive goal can be split into the two subgoals

achieve $A[i:g] \leq w$, $w \leq A[h:j]$, $i \leq g$, $h \leq j$
    varying A, g, h, w
achieve $g+1 = h$
    protecting $A[i:g] \leq w$, $w \leq A[h:j]$, $i \leq g$, $h \leq j$
    varying A, g, h, w.

The second will become a loop with exit test $g+1 = h$; the first will initialize the invariants. By reducing the ranges $[i:g]$ and $[h:j]$ to single elements, the first goal may be strengthened to

achieve $A[i] \leq w$, $w \leq A[j]$, $g = i$, $h = j$ **varying** A, g, h, w;

the first conjunct may be further strengthened to $A[i] = w$:

achieve $A[i] = w$, $w \leq A[j]$, $g = i$, $h = j$ **varying** A, g, h, w.

At this point, we would like to assign values to w, g, and h. Before we can do that we must substitute for the other occurrences of w:

achieve $A[i] = w$, $A[i] \leq A[j]$, $g = i$, $h = j$ **varying** A, g, h, w.

Splitting this into two disjoint goals and assigning we get

> **achieve** $A[i] \leq A[j]$, $\{A[i:j]\} = \{A'[i:j]\}$ **varying** A
> $(g,h,w) := (i,j,A[i])$

(with the permutation requirement included). As in the first solution, the remaining subgoal yields the conditional

> **if** $A[i] > A[j]$ **then** $(A[i],A[j]) := (A[j],A[i])$ **fi**.

The current status of the program is

B_8: **assert** $i \in N$, $j \in N$, $i < j$
if $A[i] > A[j]$ **then** $(A[i],A[j]) := (A[j],A[i])$ **fi**
$(g,h,w) := (i,j,A[i])$
assert $w = A[i]$, $A[i] \leq A[j]$, $g = i$, $h = j$
loop L_8: **assert** $A[i:g] \leq w$, $w \leq A[h:j]$, $i \leq g$, $h \leq j$
    until $g+1 = h$
    approach $g+1 = h$
               protecting $A[i:g] \leq w$, $w \leq A[h:j]$, $i \leq g$, $h \leq j$
               varying A,g,h,w
    repeat.

We may now determine bounds on g and h and apply the termination rule. Upon entering the loop $g = i < j = h$, while upon termination $i \leq g_E + 1 = h_E \leq j$. This suggests keeping $g, h \in Z$ and letting g increase from i to its final value g_E, while h decreases from j to h_E. The resulting bounds, $i \leq g \leq g_E$ and $h_E \leq h \leq j$, combined with $g_E = h_E - 1 < h_E$, imply the invariant $g < h$. So to ensure termination, we require that g and h remain integers, and that progress is made by increasing g and/or decreasing h, until they meet somewhere in the middle. Accordingly, the loop-body subgoal becomes

> **achieve** $g' \leq g$, $h' \geq h$, $g' < g \lor h' > h$
>   **protecting** $A[i:g] \leq w$, $w \leq A[h:j]$,
> $g \in Z$, $h \in Z$, $i \leq g \leq h \leq j$
>   **varying** A,g,h,w.

Splitting the two ranges into the part that has already been achieved and what remains to be achieved, we get

achieve $g' \leq g, h' \geq h, g' < g \lor h' > h,$
$\qquad A[g'+1:g] \leq w, w \leq A[h:h'-1]$
$\qquad$ **protecting** $A[i:g] \leq w, w \leq A[h:j], g \in \mathbf{Z}, h \in \mathbf{Z}, i \leq g \leq j$
$\qquad$ **varying** $A,g,h,w.$

We shall protect $A[i:g] \leq w$ and $w \leq A[h:j]$ by not varying w or any of the elements in the array segments $A[i:g]$ and $A[h:j]$. Now if we reduce the range $[g'+1:g]$ to a single element (letting $g'+1{=}g$) and make the range $[h:h'-1]$ vacuously true (insisting that $h > h'-1$), then we get

achieve $g' \leq g, h' \geq h, g' < g \lor h' > h,$
$\qquad A[g'+1] \leq w, g'+1{=}g, h > h'-1,$
$\qquad$ **protecting** $A[i:g] \leq w, w \leq A[h:j],$
$\qquad\qquad\qquad g \in \mathbf{Z}, h \in \mathbf{Z}, i \leq g < h \leq j$
$\qquad$ **varying** $A,g,h,$

which simplifies to

achieve $g'+1{=}g, h'{=}h, A[g'+1] \leq w$
$\qquad$ **protecting** $A[i:g] \leq w, w \leq A[h:j]$
$\qquad$ **varying** $A,g,h.$

The conditional rule suggests achieving the conjunct $A[g'+1] \leq w$ by testing:

if $A[g+1] \leq w$ **then** $g := g+1$
$\qquad\qquad\qquad$ **else**   **assert** $w < A[g]$
$\qquad\qquad\qquad\qquad$ **achieve** $g' \leq g, h' \geq h, g' < g \lor h' > h,$
$\qquad\qquad\qquad\qquad\qquad A[g'+1:g] \leq w, w \leq A[h:h'-1]$
$\qquad\qquad\qquad\qquad$ **protecting** $\cdots$
$\qquad\qquad\qquad\qquad$ **varying** g,h $\qquad\qquad\qquad\qquad$ **fi.**

Without going into more detail, the remaining subgoal generates two more cases: if $w \leq A[h'-1]$, then h is decremented by 1; otherwise, $A[h'-1] < w < A[g'+1]$ and, therefore, $A[h'-1]$ can be exchanged with $A[g'+1]$, increasing g and decreasing h. The completed program is

P_8: **begin comment** *partition program*
  B_8: **assert** $i \in \mathbf{N}$, $j \in \mathbf{N}$, $i < j$
  if $A[i] > A[j]$ **then** $(A[i],A[j]) := (A[j],A[i])$ **fi**
  $(g,h,w) := (i,j,A[i])$
  loop L_8: **assert** $A[i:g] \leq w$, $w \leq A[h:j]$, $i \leq g$, $h \leq j$
           until $g+1=h$
           if $A[g+1] \leq w$ **then** $g := g+1$
           else
           if $w \leq A[h-1]$ **then** $h := h-1$
                           else $(g,h) := (g+1,h-1)$
                                  $(A[g],A[h]) := (A[h],A[g])$ **fi fi**
       repeat
  E_8: **assert** $A[i:g] \leq A[g+1:j]$, $i \leq g$, $g+1 \leq j$,
           $\{A[i:j]\}=\{A'[i:j]\}$
  end

5.4. Discussion

In this chapter, we have seen how to systematically develop a program from its specification; in the next chapter we shall see how similar ideas may be employed to generate invariants from code. In general, at each stage in the synthesis of a program there is more than one unachieved subgoal, and for each subgoal there may be a number of possible rules that can be applied. Whenever a given choice turns out to be unsuccessful, a different possibility must be tried. We have not, however, addressed the issue of how one might guess which possibility to explore next. [Kant81] describes a system that guides the choices made by a synthesis system based upon analysis of time and space requirements. Annotation techniques, such as are described in the next chapter, can facilitate such analyses.

There is a wide spectrum of possibilities in program synthesis, depending on the means of specifying the goal and the kind of information used along the way. The range of specification methods include:

- *Natural language.* Possibilities and problems with "natural language" specification have been explored in, for example, [Green76], [BalzerGoldmanWile77], [Steinberg80], and [Biermann81].

- *Traces.* Developing a program from a trace of the intended execution sequence has been discussed in [BiermannBaumPetry75], [SiklossySykes75], [BiermannKrishnaswamy76], [Phillips77], [Bauer79], and [Villemin81].

- *Examples.* Specifying an algorithm by means of a set of pairs of input and output values has been investigated in [BiggerstaffJohnson77], [Summers77], [Biermann78], [JouannaudGuiho79], [JouannaudKodratoff80], [Shapiro83], and others. For a survey, see [Smith80].

- *Formal specifications.* We have considered "mathematical" specifications. Other work in synthesis taking this approach to specification includes [Darlington75], [MannaWaldinger75], [BurstallDarlington77], [MannaWaldinger79], [Bibel80], [Follett80], [MannaWaldinger80], [Scherlis80], [Clark81], [Darlington81], [Feather82], [Smith82], and [Dershowitz83].

The kinds of programming knowledge and expertise used to guide a synthesis can also take many forms. They include:

- *Expert knowledge.* Some reports of synthesis research that attempt to incorporate techniques and/or knowledge of professional programmers are [Hardy75], [ShawSwartoutGreen75], [Sussman75], [Long77], [Barstow79], [HayesKlahrMostow80], [PhillipsGreen80], and [Wood80].

- *Analogy with other programs.* The idea of developing new programs based on existing ones for different tasks, as illustrated in Chapter 3, has been suggested in [DershowitzManna77] and [UlrichMoll77].

● *Deduction.* Using formal methods to deduce programs from specifications, as we have done in this chapter, is the approach taken in [Darlington73], [DegliMiglioliOrnaghi74], [MannaWaldinger75], [BalzerGoldmanWile76], [BurstallDarlington77], [Wand77], [Blikle78], [Goto79], [Sato79], [MannaWaldinger80], [Clark81], [Hogger81], [ArsacKodratoff82], [Dershowitz83], [Paige83], and others.   [Scherlis80] and [Kott82] address the issue of ensuring the correctness of transformations used in program development.

Chapter 6

Program Annotation and Analysis

*There are ... obvious analogies between the abstraction of
ideas—the formation of concepts—and perceptual
generalization—the extraction of invariant features, stripped
of their accidental accompaniments, from varied situations.*

—*Arthur Koestler (The act of creation)*

6.1. Introduction

As we have seen, invariant assertions are often needed for
modification and abstraction to carry through. But what if the programmer failed to supply enough of them? In particular, if the program is
incorrect with respect to its specifications, then, perforce, some of his
assertions (at very least, the output specification) do not reflect what the
program is actually doing. And without knowing what the program is
doing, we cannot proceed to debug it.

In the previous chapter we saw rules for generating code from
invariants; in this chapter rules are used to derive invariants from code.
A program is given along with its specification, but it is not known

whether the program meets that specification. The task is to annotate the program with invariants describing the workings of the program as is, independent of its correctness or incorrectness. The process is iterative, since finding some invariants may suggest others. Assertions supplied by the programmer cannot be assumed true, although they may be used to guide the search for correct invariants.

If the invariants associated with the point of termination of a program imply that the given output specification is true for any input satisfying the given input specification, then the program has been proved correct. On the other hand, if there exist legal input values such that whenever the output invariants hold for those values, the output specification does not hold completely, then the program is incorrect. In this manner, invariants are used for proving correctness and incorrectness. Invariants also play important roles in other aspects of programming, including: proving termination, guiding debugging, analyzing efficiency, and aiding in optimization.

Existing implementations of the invariant-assertion method of program verification are not fully mechanical; the user must supply most, if not all, of the invariants himself. If the original program is not supplied with sufficient invariants to prove correctness or incorrectness, they must be supplemented. These invariants then enable one to verify if what the program does is what it was intended to do. Invariants are also useful in analyzing other properties of programs, such as time complexity.

In the following sections, we present a unified approach to program annotation, using annotation rules to derive invariants. Section 6.2 presents an overview of the approach. It is followed by three detailed examples in Section 6.3: the first illustrates the basic methods on a single-loop program; the second applies the methods to a program involving arrays; the third to nested loops. A catalog of forty-one annotation rules is included in Appendix 4. A description of our implementation may be found in Appendix 5.

Other implemented annotation systems include:

- the system described in [Elspas74], based mainly upon the solution of difference equations;

- VISTA ([German74], [GermanWegbreit75]), based upon the top-down heuristics of [Wegbreit74];

- ADI [Tamir80], an interactive system based upon methods of [Katz-Manna76] and [Katz76];

- [SuzukiIshihata77] and [German78] describe systems that generate invariants useful in checking for various runtime errors;

- the symbolic evaluator in [CheathamHollowayTownley79] solves difference equations for loop assignments.

6.2. Overview

In this section, we define some terminology and then present samples of the different types of annotation rules.

6.2.1. Notation and Terminology

Given a program with its specification, the goal is to document the program automatically with assertions. If the program is correct with respect to the specification, the assertions should provide sufficient information to demonstrate its correctness; if the program is incorrect, one would like information helpful in determining what is wrong.

Three types of assertions are considered:

- *Global invariants* are relations that hold at all places (i.e. labels) and at all times during the execution of some program segment. We write

$$\textbf{assert } \alpha \textbf{ in } P$$

to indicate that the relation α is a global invariant in a program segment P.[1]

---

[1]Actually, we consider α to be a global invariant in P even if it only begins to hold after the variables in α have been assigned an initial value within P.

- *Local invariants* are associated with specific points in the program, and hold for the current values of the variables whenever control passes through the corresponding point. Thus,

$$L: \textbf{assert } \alpha$$

means that the relation α holds each time control is at label L.

- *Candidate assertions*, also associated with specific points, are relations hypothesized to be local invariants, but that have not yet been verified. We write

$$L: \textbf{suggest } \alpha.$$

Consider the following simple program, meant to compute the quotient q and remainder r of the integer input values c and d:

> P_4: **begin comment** *integer–division program*
> B_4: **assert** $c \in N$, $d \in N+1$
> $(q,r) := (0,c)$
>   **loop** L_4: **assert** $\cdots$
>   **until** $r < d$
> $(q,r) := (q+1,r-d)$
>   **repeat**
> E_4: **suggest** $q \leq c/d$, $c/d < q+1$, $q \in Z$, $r = c - q \cdot d$
>   **end**

where N is the set of nonnegative integers, $N+1$ is the set of positive integers, and Z is the set of all integers. This program will be used only to illustrate various aspects of program annotation; complete examples are given in the next section.

The invariant

$$\textbf{assert } c \in N, \ d \in N+1$$

attached to the **begin**-label B_4 is the input specification of the program defining the class of "legal inputs." It indicates that whenever computation starts at B_4, the input values c and d are a nonnegative and positive integer, respectively. The input specification is assumed to hold, regardless of whether the program is correct or not.

The candidate

suggest $q \leq c/d$, $c/d < q+1$, $q \in \mathbf{Z}$, $r = c - q \cdot d$

attached to the **end**-label E_4 is the output specification of the program. It states that the desired outcome of the program is that q be the largest integer not larger than c/d and r be the remainder. Since one cannot assume that the programmer has not erred, initially all programmer-supplied assertions—including the program's output specification—are only candidates for invariants.

In order to verify that a candidate is indeed a local invariant, one must show that whenever control reaches the corresponding point, the candidate holds. Suppose that we are given a candidate for a loop invariant

L_4: **suggest** $r = c - q \cdot d$.

To prove that it is an invariant, one must show:

1) that the relation holds at L_4 when the loop is first entered, and

2) that once it holds at L_4, it remains true each subsequent time control returns to L_4.

If we succeed, then we would write

L_4: **assert** $r = c - q \cdot d$.

Furthermore, if $r = c - q \cdot d$ holds whenever control is at L_4, then it will also hold whenever control leaves the loop and reaches E_4. In other words, $r = c - q \cdot d$ would also be an invariant at E_4 and may be removed from the list of candidates at E_4. In that case, we would write

E_4: **assert** $r = c - q \cdot d$
    suggest $q \leq c/d$, $c/d < q+1$, $q \in \mathbf{Z}$

Global invariants often express the range of variables. For example, since the variable q is first initialized to 0 and then repeatedly incremented by 1, it is obvious that the value of q is always a nonnegative integer. Thus, we have the global invariant

assert $q \in \mathbf{N}$ **in** P_4

that relates to the program as a whole and states that $q \in \mathbf{N}$ throughout execution of the program segment P_4.

In this section we describe various annotation techniques. These techniques are expressed as rules: the antecedents of each rule are usually annotated program segments containing invariants or candidates and the consequent is either an invariant or a candidate. They are numbered <1>, <2>, etc. and are listed in Appendix 4. Not only are these rules useful for automatic or interactive annotation, but they help to clarify the interrelation between program text and invariants for the programmer.

We distinguish between three types of rules: assignment rules, control rules, and heuristic rules.

- *Assignment rules* yield *global* invariants based only upon the assignment statements of the program.

- *Control rules* yield *local* invariants based upon the control structure of the program.

- *Heuristic rules* have *candidates* as their consequents. These candidates, although they may be promising, are not guaranteed to be invariants.

The assignment and control rules are *algorithmic* in the sense that they derive relations in such a manner as to *guarantee* that they are invariants. The heuristics are rules of *plausible* inference, reflecting common programming practice.

6.2.2. Assignment Rules

Many of the algorithmic rules depend only upon the assignment statements of the program and not upon its control structure. In other words, whether the assignments appear, for example, within an iterative or recursive loop or on some branch of a conditional statement is irrelevant. Since the location and order in which assignments are executed does not affect the validity of the rules, these rules yield global invariants.

The various assignment rules relate to particular operators occurring in the assignment statements of the program. Some of the rules for addition, for example, are: an *addition rule* that gives the range of a variable that is updated by adding (or subtracting) a constant; a *set-addition rule* for the case where the variable is added to another variable whose range is already known; and an *addition-relation rule* that relates two variables that are always incremented by similar expressions. Corresponding rules apply to other operators.

In dealing with sets, we find the following notation convenient: let $f(s_1, s_2, \cdots, s_m)$ be any expression containing occurrences of m distinct subexpressions $s_1, s_2, \cdots, s_m$. The set of elements

$$\{ f(s_1, s_2, \cdots, s_m) : s_1 \in S_1, s_2 \in S_2, \cdots, s_m \in S_m \}$$

is denoted by[2]

$$f(S_1, S_2, \cdots, S_m).$$

The following *addition rule* uses this notation:

$<1>$ *addition rule*
$x := a_0 \mid x := x + a_1 \mid x := x + a_2 \mid \cdots$ **in** P
assert $x \in a_0 + a_1 \cdot \mathbf{N} + a_2 \cdot \mathbf{N} + \cdots$ **in** P

where P represents any program fragment and the expressions a_i are of *constant* value within P. The antecedent

$$x := a_0 \mid x := x + a_1 \mid x := x + a_2 \mid \quad \cdots \quad \textbf{in } P$$

indicates that the assignments to the variables x in P are $x := a_0$, $x := x + a_1$, $x := x + a_2$, etc. The consequent

$$\textbf{assert } x \in a_0 + a_1 \cdot \mathbf{N} + a_2 \cdot \mathbf{N} + \quad \cdots \quad \textbf{in } P$$

---

[2]This notation cannot be used when there are more than one occurrence of a subexpression s_i. For example, the set $\{ n + n : n \in \mathbf{N} \}$ is not the same as $\mathbf{N} + \mathbf{N}$ but is the same as $2 \cdot \mathbf{N}$.

is a global invariant to the effect that x belongs to the set $a_0 + a_1 \cdot N + a_2 \cdot N + \cdots$, i.e. $x = a_0 + a_1 \cdot n_1 + a_2 \cdot n_2 + \cdots$ for some $n_0, n_1, n_2, \cdots \in N$. This relation holds throughout execution of P—but only from the point when x first receives a defined value in P via the assignment $x := a_0$.[3] From such an invariant, more specific properties, such as bounds on x, may be derived. Note that no restrictions are placed on the order in which the assignments to x are executed, except that prior to the first execution of $x := a_0$ the invariant may not hold.

In our simple program P_4, the assignments to the variable q are

$$q := 0 \mid q := q + 1.$$

So we can apply the *addition rule,* instantiating a_0 with 0 and a_1 with 1, and obtain the global invariant $q \in 0 + 1 \cdot N$, i.e.

assert $q \in N$ **in** P_4.

The assignments to r in P_4 are

$$r := c \mid r := r - d.$$

Applying the same rule to them, letting $a_0 = c$ and $a_1 = -d$, yields the invariant

assert $r \in c - d \cdot N$ **in** P_4.

Given that d is positive, one may conclude that $r \leq c$.

The *set-addition rule* is a more general form of the above *addition rule,* applicable to nondeterministic assignments of the form $x :\in f(S)$, whereby an arbitrary element in the set $f(S) = \{f(s) : s \in S\}$ is assigned to x. In general, for the set rules, an assignment $x := f(s)$, where it is only known that $s \in S$, may be viewed as a nondeterministic assignment $x :\in f(S)$. The *set-addition rule* is

---

[3]After any execution of $x := a_0$, clearly $x \in a_0 + a_1 \cdot N + a_2 \cdot N + \cdots$ with $x = a_0 + a_1 \cdot 0 + a_2 \cdot 0 + \cdots$. If $x = a_0 + a_1 \cdot n_0 + a_2 \cdot n_1 + \cdots$ for some $n_0, n_1, \ldots$ before executing $x := x + a_1$, then $x = a_0 + a_1 \cdot (n_0 + 1) + a_2 \cdot n_1 + \cdots$ after executing the assignment. Thus, n_0 represents the number of executions of $x := x + a_1$ since $x := a_0$ was executed last, n_1 is the number of executions of $x := x + a_2$, etc.

```
| <5>  set-addition rule                                  |
| x :∈ S₀ | x :∈ x+ S₁ | x :∈ x+ S₂ |  · · ·  in P        |
| assert x ∈ S₀+ ΣS₁+ ΣS₂+ · · ·  in P                    |
```

where ΣS denotes the set of finite sums $s_1+ s_2+ \cdots + s_m$ for (not necessarily distinct) addends s_i in S.[4] (If $S=\{\}$, the sum is 0; if S contains the single element s, then $\Sigma S = s \cdot \mathbf{N}$.) In program P_4 the assignments to r were

$$r := c \mid r := r - d.$$

Since we are given that $c \in \mathbf{N}$ and $d \in \mathbf{N}+1$, we may consider the non-deterministic assignments

$$r :\in \mathbf{N} \mid r :\in r - (\mathbf{N}+1),$$

and by applying the *set-addition rule* we obtain the global invariant $r \in \mathbf{N} - \Sigma(\mathbf{N}+1)$. This simplifies to

$$\textbf{assert } r \in \mathbf{Z} \textbf{ in } P_4,$$

where $\mathbf{Z}$ is the set of all integers.

To relate different variables appearing in a program, there is an *addition-relation rule*

```
| <11>  addition-relation rule                            |
| (x,y) := (a₀,b₀)                                        |
|  | (x,y) := (x+ a₁·u, y+ b₁·u)                          |
|  | (x,y) := (x+ a₁·v, y+ b₁·v) |  · · ·  in P           |
| assert a₁·(y − b₀)= b₁·(x − a₀)  in P                   |
```

where u, v, ..., are arbitrary (not necessarily constant) expressions. The invariant begins to hold only when the multiple assignment $(x,y) := (a_0, b_0)$ has been executed for the first time.[5] The multiple

---

[4]This rule applies analogously to any associative and commutative operator $\oplus$.

[5]The invariant clearly holds when $x = a_0$ and $y = b_0$. Assuming it holds before executing $(x,y) := (x + a_1 \cdot u, y + b_1 \cdot u)$, then after executing the assignment, both sides of the equality are increased by $a_1 \cdot b_1 \cdot u$, and the invariant still holds.

assignments in the antecedent of such rules, e.g. $(x,y) := (x + a_1 \cdot u, y + b_1 \cdot u)$, may represent the cumulative effect of individual assignments lying on a path between two labels, with the understanding that whenever $x := x + a_1 \cdot u$ is executed, so is $y := y + b_1 \cdot u$ for the same value of the expression u. In that case, the invariant will not, in general, hold between the individual assignments. Note that this *addition-relation rule* (as well as several other relation rules) may be derived from the following general relation-rule schema:

$$
\begin{array}{|l|}
\hline
(x,y) := (a_0, b_0) \\
\mid (x,y) := (x \oplus (u \otimes a_1), y \oplus (u \otimes b_1)) \\
\mid (x,y) := (x \oplus (v \otimes a_1), y \oplus (v \otimes b_1)) \mid \; \cdots \; \textbf{in } P \\
\hline
\textbf{assert } (a_0 \otimes b_1) \oplus (y \otimes a_1) = (b_0 \otimes a_1) \oplus (x \otimes b_1) \textbf{ in } P \\
\hline
\end{array}
$$

where the operator $\oplus$ is commutative and associative, i.e. $u \oplus v = v \oplus u$ and $u \oplus (v \oplus w) = (u \oplus v) \oplus w$, the operator $\otimes$ is right-permutative, i.e. $(u \otimes v) \otimes w = (u \otimes w) \otimes v$, and $\otimes$ distributes over $\oplus$, i.e. $(u \oplus v) \otimes w = (u \otimes w) \oplus (v \otimes w)$.

In our example, the assignments in the initialization path give us

$$(q,r) := (0, c),$$

and for the loop-body path we have

$$(q,r) := (q + 1, r - d).$$

By a simple application of the *addition-relation rule* with $a_0 = 0$, $b_0 = c$, $a_1 = v = 1$, and $b_1 = -d$, we derive the invariant $1 \cdot (r - c) = -d \cdot (q - 0)$, which simplifies to

$$\textbf{assert } r = c - q \cdot d \textbf{ in } P_4.$$

The assignment rules for range invariants are related to the weak interpretation method of [Sintzoff72] (see also [Wegbreit75], [WegbreitSpitzen76], and [Harrison77]; for implementations see [Scherlis74] and [GermanWegbreit75].) The relation rules are related to optimization techniques, to the "metatheorems" in [MannaPnueli74], and to the approach in [Caplain75]. [German78] and [Ellozy81] apply similar methods. In the previous chapter we saw how such rules may also be used to extend a program to achieve additional relations.

6.2.3. Control Rules

Unlike the previous rules that completely ignore the control structure of the program, control rules derive important invariants from the program structure; they are akin to the verification rules of [Hoare69]. There are several rules for each program construct, for example, two rules push invariants forward in a loop.

The *forward loop-exit rule*

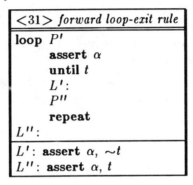

reflects the fact that if execution of a loop terminates at L'', then the exit test t must have just held, while if the loop is continued at L', the exit test was false. Furthermore, any relation α that held just prior to the test, also holds immediately after. Applying this rule to the loop in the integer-division program P_4 yields the invariant $r < d$ at E_4 and $r \geq d$ at the head of the loop body:

$$\textbf{assert } r \geq d$$
$$(q,r) := (q+1, r-d)$$

To propagate invariants such as $r \geq d$ past assignment statements, there is a *forward assignment rule,*

$<21a>$ *forward assignment rule*
assert $\alpha(x,y)$ $x := f(x,y)$ $L:$
$L:$ **assert** $\alpha(f^-(x,y),y)$

where f^- is the inverse (assuming that there is an inverse) of the function f in the first argument, i.e. $f^-(f(x,y),y)=x$. By using the inverse function f^-, the value of x prior to the assignment may be expressed in terms of the current value of x as $f^-(x,y)$. Thus, if the relation $\alpha(x,y)$ held before the assignment to x, then after the assignment $\alpha(f^-(x,y),y)$ holds, where $f^-(x,y)$ has been substituted for all occurrences of x in $\alpha(x,y)$.[6] In our example, since the first loop-body assignment $q := q+1$ does not affect any variable appearing in the invariant $r \geq d$, the invariant is pushed forward unchanged. To propagate $r \geq d$ past the second assignment, $r := r - d$, we replace r by the inverse of $r-d$, which is $r+d$, yielding $r+d \geq d$, or

$$\textbf{assert } r \geq 0,$$

at the end of the loop body. There is also an axiom for assignments

$<18>$ *assignment axiom*
$x := a$ **assert** $x=a$

where the expression a may not contain x. This axiom gives the invariant

$$\textbf{assert } r=c$$

prior to entering the loop.

---

[6]Such use of inverse functions was suggested in [Elspas74]. Even if there is no inverse function, variants of this rule may often be used to glean some useful information.

The rule

$<29>$ *forward loop-body rule*
assert α **loop** L: $\quad P$ $\quad\quad$**assert** β $\quad$**repeat**
L: **assert** $\alpha \vee \beta$

states that for control to be at the head of a loop, at L, either the loop
has just been entered, or the loop body has been executed and the loop is
being repeated. Therefore, the disjunction $\alpha \vee \beta$ of an invariant α
known to hold just before the loop and an invariant β known to hold at
the end of the loop body must hold at L. Using this second rule for
loops, we get the loop invariant

$$L_4: \textbf{assert } r=c \vee r \geq 0$$

By the input specification $c \in \mathbf{N}$, the first disjunct $r=c$ implies the
second $r \geq 0$. Therefore, this invariant simplifies to

$$L_4: \textbf{assert } r \geq 0.$$

With the global invariant $r=c-q\cdot d$ and input specification $d \in \mathbf{N}+1$, it
follows that $q \leq c/d$.

6.2.4. Schematic Rules

In Chapter 4, we saw a method for deriving program schemata and
applying them. In this subsection, we illustrate how control rules may be
used to annotate schemata. Once invariants have been generated for a
particular schema, they can be used for all its instances. Related work,
making use of a small collection of "loop-plans" to decompose program
loops, may be found in [Waters77], where it is claimed that the
overwhelming majority of loops in software packages such as [IBM70] are
easily analyzed.

Consider, for example, the following single-loop, single-conditional, program schema:

```
S*: begin comment single–loop schema
    z := c
    loop L*: assert  · · ·
             until t(z)
             z := f(z)
             if s(z) then z := g(z) else z := h(z) fi
             repeat
    end
```

The *assignment axiom* $\langle 18 \rangle$, when applied to the initial assignment $z := c$, yields the invariant

$$\textbf{assert } z = c$$

before the loop. The

$\langle 33 \rangle$ *label axiom*
$L:$ **assert** $x = x_L$

where x_L is the value of x when last at L, yields $z = z_L$. at the head of the loop body of the schema, preceding the **until**-clause. This rule allows for relating values of a variable at different points within a program. The *forward loop-exit rule* $\langle 31 \rangle$ generates the invariant $\sim t(z)$ after the test, from which it follows that $\sim t(z_L \cdot)$. Then the

$\langle 21c \rangle$ *simple forward assignment rule*
assert $x = a$ $x := f(x,y)$ $L:$
$L:$ **assert** $x = f(a,y)$

where a must be a constant expression, gives $z = f(z_L \cdot)$ after the assignment and before the conditional:

$$\textbf{assert } \sim t(z_L \cdot), \; z = f(z_L \cdot)$$
$$\textbf{if } s(z) \textbf{ then } z := g(z) \textbf{ else } z := h(z) \textbf{ fi.}$$

(The invariant $\sim t(z_L.)$ is propagated forward unchanged by rule $<21a>$.)

To generate invariants from a conditional test, we have a *forward test rule:*

$<25>$ *forward test rule*
assert α **if** t **then** $\quad L'$: $\qquad\qquad\qquad P'$ $\qquad\quad$ **else** $\quad L''$: $\qquad\qquad\qquad P''\qquad$ **fi**
L': **assert** α, t L'': **assert** $\alpha, \sim t$

That is, for the **then**-branch to be taken t must be true, while for the **else**-branch to be taken it must be false; furthermore, any α that held before the test, also holds after. Once invariants have been generated for the two branches, they are pushed forward by the *forward branch rule*

$<27>$ *forward branch rule*
if t $\quad$ **then** $\quad P'$ $\qquad\qquad\qquad$ **assert** α $\qquad\quad$ **else** $\quad P''$ $\qquad\qquad\qquad$ **assert** $\beta\quad$ **fi** L:
L: **assert** $\alpha \vee \beta$

It states that for control to be at the point after the conditional statement, one of the two branches must have been traversed.

For our schema, the *forward test rule* $<25>$ propagates the invariants forward and adds $s(z)$—and hence $s(f(z_L.))$—to the head of the **then**-clause of the conditional, as well as $\sim s(f(z_L.))$ at the head of the **else**-clause:

> **if** $s(z)$ **then**   **assert** $\sim t(z_L\cdot)$, $s(f(z_L\cdot))$, $z = f(z_L\cdot)$
> $\qquad\qquad\qquad z := g(z)$
> $\qquad$ **else**   **assert** $\sim t(z_L\cdot)$, $\sim s(f(z_L\cdot))$, $z = f(z_L\cdot)$
> $\qquad\qquad\qquad z := h(z)$ $\qquad\qquad\qquad\qquad\qquad$ **fi**

In a similar manner, pushing these invariants through the **then**- and **else**- branch assignments, we get

> **if** $s(z)$ **then** $z := g(z)$
> $\qquad\qquad\quad$ **assert** $\sim t(z_L\cdot)$, $s(f(z_L\cdot))$, $z = g(f(z_L\cdot))$
> $\qquad$ **else** $z := h(z)$
> $\qquad\qquad\quad$ **assert** $\sim t(z_L\cdot)$, $\sim s(f(z_L\cdot))$, $z = h(f(z_L\cdot))$ **fi**.

Combining the invariants from the two different paths—using the *forward branch rule* $<27>$—one obtains

$$\textbf{assert } \sim t(z_L\cdot),\ [s(f(z_L\cdot)) \wedge z = g(f(z_L\cdot))]$$
$$\vee [\sim s(f(z_L\cdot)) \wedge z = h(f(z_L\cdot))]$$

after the conditional, at the end of the loop body.

When the inverse functions f^-, g^-, and h^- are available, the $z_L\cdot$ may be eliminated, giving instead

$$\textbf{assert } [\sim t(f^-(g^-(z))) \wedge s(g^-(z))] \vee [\sim t(f^-(h^-(z))) \wedge \sim s(h^-(z))]$$

In that case, we may apply the *forward loop-body rule* $<29>$ to the schema, and derive the loop invariant

L^*: **assert** $z = c \vee [\sim t(f^-(g^-(z)) \wedge s(g^-(z))] \vee [\sim t(f^-(h^-(z))) \wedge \sim s(h^-(z))]$.

This loop invariant embodies two facts about the control structure of the schema.

● Whenever control is at L^*, either the loop has just been entered or the loop-exit test was false the last time around the loop. That is,

$\qquad L^*$: **assert** $z = c \vee \sim t(f^-(g^-(z))) \vee \sim t(f^-(h^-(z)))$.

The first disjunct is the result of the initialization path; the second states that the exit test was false for the value of z when L^* was last visited, assuming control came via the **then**-path of the conditional; the third disjunct says the same for the case when control came via the **else**-path.

- Whenever control is at L^*, either the loop has just been entered, or the conditional test was true the last time around and the **then**-path was taken, or the test was false and the **else**-path was taken. That is,

$$L^*: \text{ assert } z = c \lor s(g^-(z)) \lor \sim s(h^-(z)).$$

The following rule is valuable for programs with universally-quantified output specification. Given a loop invariant $\alpha(x)$ at L containing the integer variable (or expression) x and no other variables, check if x is monotonically increasing by one. If it is, then we have as a loop invariant at L that α still holds for all intermediate values lying between the initial and current values. That is

---

$<35>$ *forall rule*

---

assert $x = a$, $x \in \mathbf{Z}$
loop L: **assert** $\alpha(x)$
    P
    assert $x = x_L + 1$
    repeat

L: **assert** $(\forall \xi \in [a:x]) \; \alpha(\xi)$

---

where a is an integer expression with a constant value in P and $[a:x]$ is the set of integers between a and x, inclusive.[7]

As a simple example, consider the loop schema

S_3: **begin** *linear−search schema*
    $z := 0$
    loop L_3: **assert** $\cdots$
        until $t(z)$
        $z := z + 1$
        repeat
    E_3: **assert** $\cdots$
    end

By the *label axiom* $<33>$, we have

---

[7]This rule is similar to the universal-quantification technique for arrays in [Katz-Manna73].

$$L_3: \text{ assert } z = z_{L_3}.$$

Thus, we can easily derive the following invariants:

$z := 0$
assert $z = 0$
loop $L_3:$ **assert** $z = z_{L_3}$
   until $t(z)$
   $z := z + 1$
   assert $z - 1 = z_{L_3}, \ \sim t(z-1)$
   repeat
$E_3:$ **assert** $t(z)$

Now, by the *forward loop-body rule* <29> we can derive the invariant

$$L_3: \text{ assert } z = 0 \lor \sim t(z-1)$$

and by the *forall rule*, we get

$$L_3: \text{ assert } (\forall \xi \in [0:z]) \ [\xi = 0 \lor \sim t(\xi - 1)].$$

This simplifies to

$$L_3: \text{ assert } (\forall \xi \in [0:z-1]) \ \sim t(\xi).$$

Combined with the invariant $t(z)$ that holds at E_3, it implies that the final value of z is the minimum nonnegative integer satisfying the predicate t, i.e.

$$E_3: \text{ assert } z = \min_{\xi \in \mathbf{N}} t(\xi).$$

6.2.5. Heuristic Rules

In contrast with the above rules that derive relations that are guaranteed to be invariants, there is another class of rules, heuristic rules, that can only suggest candidates for invariants. Candidates must be verified.

As an example, consider the following *conditional heuristic*

$<36>$ *conditional heuristic*
if t **then** P'
assert α
else P''
assert β **fi**
$L:$
$L:$ **suggest** $\alpha,\ \beta$

Since we know that α holds if the **then**-path P' is taken, while β holds if the **else**-path P'' is taken, clearly their disjunction $\alpha \lor \beta$ holds at L in either case (that was expressed in the *forward branch rule* $<27>$). However, since in constructing a program, a conditional statement is often used to achieve the same relation in alternative cases,[8] it is plausible that α (or, by the same token, β) may hold true for *both* paths.

As mentioned earlier, the output specification and user-supplied assertions are the initial set of candidates. Candidates are propagated over assignment and conditional statements using the same control rules as for invariants. Furthermore, one candidate may suggest another. The *top-down heuristic,*

$<38>$ *top-down heuristic*
suggest α
loop $L:$
until t
P
repeat
suggest γ
fact γ **when** α
$L:$ **suggest** γ

---

may be used to push a candidate (or invariant) γ backwards into a loop. Although $t \supset \gamma$ (i.e. $\sim t \lor \gamma$) would be a sufficiently strong loop invariant at L to establish γ upon loop exit, the heuristic suggests a stronger

---

[8]See Section 5.2.5.

candidate, γ itself, at L. Since a necessary condition for γ to be an invariant is that it hold upon entrance to the loop, the second antecedent of the rule requires there is reason to believe that γ holds before the loop. The idea underlying this heuristic is that an iterative loop is constructed in order to achieve a conjunctive goal $(t \wedge \gamma)$ by placing one conjunct of the goal (t) in the exit test, and maintaining the other (γ) invariantly true.[9]

In [Wegbreit74] and [KatzManna76] a more general form of the above two heuristics is suggested:

$<39>$ *disjunction heuristic*
L: **assert** $\alpha \vee \beta$
L: **suggest** $\alpha,\ \beta$

However, as remarked there, this heuristic should not be applied indiscriminately to any disjunctive invariant. One would not, for example, want to replace all occurrences of an invariant $x \geq 0$ with candidates $x > 0$ and $x = 0$. Special cases, such as the *conditional* and *top-down* heuristics are needed to indicate where the disjunctive strategy is relatively likely to be profitable.

Returning to our integer-division example P_4 the *top-down heuristic* suggests that of the candidates

$$E_4: \textbf{suggest } q \leq c/d,\ c/d < q+1,\ q \in \mathbf{Z},\ r = c - q \cdot d,$$

those that hold upon entering the loop—when $q = 0$ and $r = c$ — are also candidates at L_4. They are

$$L_4: \textbf{suggest } q \leq c/d,\ q \in \mathbf{Z},\ r = c - q \cdot d.$$

The remaining candidate at E_4, $c/d < q+1$, does not necessarily hold for $q = 0$.

Each candidate must be checked for invariance: it must hold for the loop-initialization path and must be maintained true around the loop. Of the three candidates at L_4, the last two, $q \in \mathbf{Z}$ and $r = c - q \cdot d$, have already been shown to be global invariants. To prove that the first,

---

[9]See Section 5.2.6.

$q \leq c/d$, is a loop invariant at L_4, we try to show that if $q \leq c/d$ is true at L_4 and the loop is continued, then $q \leq c/d$ holds when control returns to L_4, i.e.

$$q \leq c/d \wedge r \geq d \supset q+1 \leq c/d.$$

This condition, however, is not provable. Nevertheless, we can show that $q \leq c/d$ is an invariant by making use of the global invariant $r = c - q \cdot d$. Substituting $c - q \cdot d$ for r in $r \geq d$ yields $c - q \cdot d \geq d$; it follows that the above implication holds and $q \leq c/d$ is an invariant at L_4. Thus, while an attempt to directly verify the candidate $q \leq c/d$ failed, once we have established that $r = c - q \cdot d$ is an invariant, we can also show that $q \leq c/d$ is an invariant.

Indeed, in general there may be insufficient information to prove that a candidate is invariant when it is first suggested, and only when other invariants are subsequently discovered might it become possible to verify the candidate. Therefore, candidates should be retained until all invariants and candidates have been generated. Unproved candidates may also used by the heuristics to generate additional candidates. For example, the top-down heuristic uses the as yet unproved candidate γ to generate the loop candidate γ at L.

Another heuristic, valuable for loops with universally quantified exit invariants, is

$<37>$ *generalization heuristic*
assert $x = a$ **loop** L: **suggest** $\alpha(x,y)$ $\quad\quad P$ $\quad\quad$ **assert** $x = f(x_L)$ $\quad\quad$ **repeat**
L: **suggest** $(\forall \xi \in \{a, f(a), f(f(a)), \cdots, x\})\ \alpha(\xi, y)$

Given a loop candidate $\alpha(x,y)$, we determine the set of values that the variable x takes on. Then we have as a new candidate for a loop invariant that α still holds for all its intermittent values, between the initial value a and the current value x. For example, if $a \in \mathbf{Z}$ and $f(x) = x + 1$, then as a candidate we have

$$L: \textbf{suggest} \ (\forall \, \xi \in [a:x]) \ \alpha(\xi, y).$$

Unlike the *forall rule* $<35>$, this is a candidate and not an invariant since the program segment P may vary the value of y in such a way as to destroy the relation $\alpha(x, y)$ for previous values of x.

Note that a candidate invariant must sometimes be replaced by a stronger candidate in order to prove invariance. This is analogous to other forms of proof by induction, where it is often necessary to strengthen the desired theorem to carry out a proof. The reason is that by strengthening the theorem to be proved, one is at the same time strengthening the hypothesis that is used in the inductive step. We could not, for example, directly prove that the relation $(r \geq d) \lor (r = c - q \cdot d)$ is a loop invariant (that is the necessary condition for $r = c - q \cdot d$ to hold after the loop), since this candidate is not preserved by the loop, i.e.

$$[r \geq d \lor r = c - q \cdot d] \land r \geq d \supset [r - d \geq d \lor r - d = c - (q + 1) \cdot d]$$

is not provable. On the other hand, we can prove that the stronger relation $r = c - q \cdot d$ is an invariant, since we have a stronger hypothesis on the left-hand side of the implication; that is,

$$r = c - q \cdot d \land r \geq d \supset r - d = c - (q + 1) \cdot d$$

can be proved. Clearly, once we establish that $r = c - q \cdot d$ is an invariant, it follows that $(r \geq d) \lor (r = c - q \cdot d)$ also is.

Various specific methods of strengthening candidates have been discussed in the literature ([Moriconi74], [Wegbreit74], [KatzManna76], and others). Related techniques are used in [GreifWaldinger74], [Gerhart75c], and [SuzukiIshihata77]. Also the candidates derived by the methods of [BasuMisra75], [Misra77], and [MorrisWegbreit77], using the "subgoal-induction" method of verification, fall into this class.

6.2.6. Counters

A useful method for proving certain properties of programs is augmentation with counters of various sorts. For example, by initializing a counter to zero upon entering a loop and incrementing it by one with each iteration, the value of the counter will indicate the number of times the loop has been executed. Then, relations between the program variables and the counter can be found. By deriving bounds on the counter, the termination of the loop may be proved and time complexity analyzed.

As a simple example, reconsider our (now annotated) division program

---

assert $c \in N$, $d \in N+1$, $q \in N$, $r = c - q \cdot d$ **in**
P_4: **begin comment** *annotated integer-division program*
   B_4: **assert** $c \in N$, $d \in N+1$
   $(q,r) := (0,c)$
   loop L_4: **assert** $q \leq c/d$
      until $r < d$
      $(q,r) := (q+1, r-d)$
      repeat
   E_4: **assert** $r < d$, $q \leq c/d$
   end

The variable q is incremented by 1 with each loop iteration and is initialized to 0; thus it counts iterations. Since the loop invariant $q \leq c/d$ gives an upper bound on the value of that counter, the loop must terminate. Since the output invariant $r < d$ and global invariant $r = c - q \cdot d$ yield a lower bound on the value of the counter, one can determine the exact number of loop iterations.

Examples of the use of counters for proving termination have appeared in [Knuth68], [KatzManna75b], [LuckhamSuzuki77] and [Hochhauser78]. Loop counters may also be used to discover relations between variables by solving first-order difference equations, as suggested in [Elspas,*etal.*72], [Elspas74], [GreifWaldinger74], [CheathamTownley76], and [KatzManna76].[10] (In [Netzer76] the same technique is applied to recursive programs; [CousotCousot77] generalizes the method.)

---

[10]Appendix 4 contains "counter relation" rules for this purpose.

6.3. Examples

In this section, we demonstrate how three programs taken from the program annotation literature can be annotated using the annotation rules. Our first example is the annotation of a program intended to divide two real numbers. As we shall discover, this program is not correct.[11] Appendix 5 contains a trace of the automatic annotation of the corrected version. The second example is a search for the minimum of an array, the same program synthesized in the previous chapter. The third program is designed to sort an array, and incorporates the second.

6.3.1. Real Division

Consider the following program T_1^+ purporting to approximate the quotient c/d of two nonnegative real numbers c and d, where $c < d$. Upon termination, the variable q should be no greater than the exact quotient and the difference between q and the quotient must be less than a given positive tolerance e. The program, with its input and output specifications included as assertions, is

```
T₁⁺ : begin comment real–division program
  B₁⁺ : assert 0 ≤ c < d, e > 0
  (q,qq,s,ss) := (0,0,1,d)
  loop L₁⁺ : assert  · · ·
       until s ≤ e
       if qq + ss ≤ c then (q,qq) := (q + s,qq + ss) fi
       (s,ss) := (s/2,ss/2)
       repeat
  E₁⁺ : suggest q ≤ c/d, c/d < q + e
  end
```

Our goal is to find loop invariants at L_1^+ in order to verify the output candidates at E_1^+. In our presentation of the annotation of this program, we first apply the assignment rules and then the control rules combined with one heuristic rule.

---

[11]It is a more complicated version of the program debugged in Chapter 2, containing the improvements suggested in Section 5.2.8. See also [KatzManna75a].

6.3.1.1. Assignment Rules

As a first step, one can derive simple invariants by ignoring the control structure of the program and considering only the assignment statements. This yields global invariants that hold throughout execution.

We first look for range invariants by considering all assignments to each variable. For example, since the assignments to s are

$$s := 1 \mid s := s/2,$$

we can apply the *multiplication rule*

$<2>$ *multiplication rule*
$x := a_0 \mid x := x \cdot a_1 \mid x := x \cdot a_2 \mid \ \cdots \ $ **in** P
assert $x \in a_0 \cdot a_1^N \cdot a_2^N \cdots$ **in** P

Taking 1 for a_0 and $1/2$ for a_1, we derive the global invariant

$$\textbf{assert } s \in 1/2^N \textbf{ in } T_1^+ . \tag{1}$$

In other words, $s = 1/2^n$ for some nonnegative integer n. From this it is possible to derive lower and upper bounds on s, i.e. $0 < s \leq 1$, since $s = 1$ when $n = 0$, while $s = 1/2^n$ approaches 0 as n grows larger.

Similarly, applying the *multiplication rule* to the assignments to ss,

$$ss := d \mid ss := ss/2,$$

yields

$$\textbf{assert } ss \in d/2^N \textbf{ in } T_1^+ . \tag{2}$$

Since we are given that $d > 0$, it follows that $0 < ss \leq d$.

The assignments to q are

$$q := 0 \mid q := q + s.$$

Since we know (1) $s \in 1/2^N$, these assignments may be viewed as nondeterministic assignments

$$q :\in 0 \mid q :\in q + 1/2^N.$$

Using the *set-addition rule*

> **$<5>$** *set-addition rule*
>
> $x := S_0 \mid x :\in x + S_1 \mid x :\in x + S_2 \mid \;\cdots\;$ **in** P
>
> **assert** $x \in S_0 + \Sigma S_1 + \Sigma S_2 + \;\cdots\;$ **in** P

we conclude

$$\textbf{assert } q \in \Sigma 1/2^N \textbf{ in } T_1^+ .$$

This invariant states that q is a finite sum of elements of the form $1/2^n$, where n is some nonnegative integer. Since for any two such elements, one is a multiple of the other, it follows that the sum is of the form $m/2^n$, where $m, n \in \mathbf{N}$:

$$\textbf{assert } q \in \mathbf{N}/2^N \textbf{ in } T_1^+ \tag{3}$$

(i.e. q is a dyadic rational number).

From (2) $ss \in d/2^N$ and the assignments

$$qq := 0 \mid qq := qq + ss,$$

we get $qq \in d \cdot \Sigma 1/2^N$ by the same *set-addition rule*, that is,

$$\textbf{assert } qq \in d \cdot \mathbf{N}/2^N \textbf{ in } T_1^+ . \tag{4}$$

The above four invariants (1-4) give the range of each of the four program variables. Next we take up relations between pairs of variables by considering their respective assignments. Consider, first, the variables s and ss. Their assignments are

$$(s, ss) := (1, d) \mid (s, ss) := (s/2, ss/2).$$

Each time one is halved, so is the other; therefore, the proportion between the initial values of s and ss is maintained throughout loop execution. This is an instance of the rule

> **$<12>$** *multiplication-relation rule*
>
> $(x, y) := (a_0, b_0)$
> $\mid (x, y) := (x \cdot u^{a_1}, y \cdot u^{b_1})$
> $\mid (x, y) := (x \cdot v^{a_1}, y \cdot v^{b_1}) \mid \;\cdots\;$ **in** P
>
> **assert** $x^{b_1} \cdot b_0^{a_1} = a_0^{b_1} \cdot y^{a_1}$ **in** P

yielding $s^1 \cdot d^1 = 1^1 \cdot ss^1$ which simplifies to

$$\textbf{assert } ss = d \cdot s \textbf{ in } T_1^+ . \tag{5}$$

Rules may be matched with assignments in the following manner: the pattern

$$(x,y) := (a_0, b_0)$$

matches the assignment

$$(s, ss) := (1, d)$$

by instantiating x, y, a_0, and b_0 with s, ss, 1, and d, respectively. Substituting these values in the second assignment of the rule yields

$$(s, ss) := (s \cdot u^{a_1}, ss \cdot u^{b_1})$$

which must be matched with $(s, ss) := (s/2, ss/2)$. To match $s \cdot u^{a_1} \Rightarrow s/2$, divide both sides by s, leaving $u^{a_1} \Rightarrow 1/2$. This, in turn, is effected by letting $a_1 \Rightarrow 1$ and $u \Rightarrow 1/2$. Finally, matching $ss \cdot (1/2)^{b_1} \Rightarrow ss/2$, we get $b \Rightarrow 1$. (See Appendix 1.)

The assignments to q and qq are

$$(q, qq) := (0,0) \mid (q, qq) := (q + s, qq + ss).$$

Using (5) $ss = d \cdot s$ to substitute for ss in the assignment $qq := qq + ss$, we have

$$(q, qq) := (0,0) \mid (q, qq) := (q + s, qq + d \cdot s),$$

which is an instance of the

$<11>$ *addition-relation rule*
$(x,y) := (a_0, b_0)$ $\mid (x,y) := (x + a_1 \cdot u, y + b_1 \cdot u)$ $\mid (x,y) := (x + a_1 \cdot v, y + b_1 \cdot v) \mid \;\; \cdots \;\; \textbf{in } P$
$\textbf{assert } a_1 \cdot (y - b_0) = b_1 \cdot (x - a_0) \textbf{ in } P$

Thus, we have the global invariant $1 \cdot (qq - 0) = d \cdot (q - 0)$, i.e.

$$\textbf{assert } qq = d \cdot q \textbf{ in } T_1^+ . \tag{6}$$

In all, we have established the following global invariants:

> **assert** $s \in 1/2^N$, $ss \in d/2^N$, $q \in N/2^N$,
> $qq \in d \cdot N/2^N$, $ss = d \cdot s$, $qq = d \cdot q$ **in** T_1^+.

6.3.1.2. Control Rules

So far we have derived global invariants from the assignment statements, ignoring the control structure of the program. We turn now to local invariants extracted from the program structure.

By applying the *assignment axiom*

$<18>$ *assignment axiom*
$x := a$ **assert** $x = a$

to the multiple assignment at the beginning of the program we get the local invariant

> **assert** $q = 0$, $qq = 0$, $s = 1$, $ss = d$

just prior to the loop. The axiom

$<20>$ *loop axiom*

loop P'
   until t
   assert $\sim t$
   P''
   repeat
 assert t

yields $s > e$ at the head of the loop body and $s \le e$ at E_1^+. Thus far we have the annotated program fragment

assert $q{=}0$, $qq{=}0$, $s{=}1$, $ss{=}d$
loop L_1^+ : **assert** $\cdots$
    until $s \le e$
    assert $s > e$
    if $qq + ss \le c$ **then** $(q,qq) := (q+s,qq+ss)$ **fi**
    $(s,ss) := (s/2,ss/2)$
    repeat
E_1^+ : **assert** $s \le e$

Applying the rule

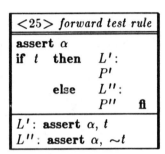

to the conditional statement of the loop[12]

$$\textbf{if } qq + ss \le c \textbf{ then } (q,qq) := (q+s,qq+ss) \textbf{ fi}$$

yields

$$\textbf{if } qq + ss \le c \quad \textbf{then} \quad \textbf{assert } s > e,\ qq + ss \le c$$
$$(q,qq) := (q+s,qq+ss)$$
$$\textbf{else} \quad \textbf{assert } s > e,\ c < qq + ss \quad \textbf{fi}$$

A variant of the *forward assignment rule* is

$<$21b$>$ *forward assignment rule*
assert $\alpha(\bar{u},\bar{y})$ $\bar{x} := \bar{u}$ $L:$
$L:$ **assert** $\alpha(\bar{x},\bar{y})$

where $\alpha(\bar{x},\bar{y})$ contains occurrences of the variables $\bar{x}$ that are on the

---

[12]This conditional is imagined to have an "empty" **else**-branch.

left-hand side of the assignment, along with other variables $\bar{y}$, and $\alpha(\bar{x},\bar{y})$ has the corresponding right-hand sides $\bar{x}$ substituted for the $\bar{x}$. Using this rule, the assignment of the **then**-branch transforms the invariant $qq + ss \leq c$ into $qq \leq c$ and leaves $s > e$ unchanged. We obtain

$$\textbf{if } qq + ss \leq c \quad \textbf{then} \quad (q,qq) := (q + s, qq + ss)$$
$$\textbf{assert } s > e, \; qq \leq c$$
$$\textbf{else} \quad \textbf{assert } s > e, \; c < qq + ss \quad \textbf{fi}$$

We may now apply the rule

$<27>$ *forward branch rule*
if t **then** P' **assert** α **else** P'' **assert** β **fi** $L:$
$L:$ **assert** $\alpha \lor \beta$

to the two possible outcomes of the conditional. We obtain the invariant

$$\textbf{assert } (s > e \land qq \leq c) \lor (s > e \land c < qq + ss),$$

which simplifies to just

$$\textbf{assert } s > e,$$

since $s > e$ appears in both disjuncts while $qq \leq c \lor c < qq + ss$ is implied by the global invariant (2) $ss > 0$.[13]

By application of the rule

$<21a>$ *forward assignment rule*
assert $\alpha(x,y)$ $x := f(x,y)$ $L:$
$L:$ **assert** $\alpha(f^-(x,y),y)$

---

[13]The disjunction $qq \leq c \lor c < qq$ is a tautology, and if ss is positive, then $c < qq$ implies $c < qq + ss$.

to the invariant $s > e$, we get

$$\textbf{assert } 2 \cdot s > e$$

at the end of the loop. By applying the rule

$<29>$ *forward loop-body rule*
assert α **loop** L: $\qquad P$ $\qquad$ **assert** β $\qquad$ **repeat**
L: **assert** $\alpha \lor \beta$

taking $s = 1$ for α and $2 \cdot s > e$ for β, we derive the loop invariant

$$L_1^+ : \textbf{assert } s = 1 \lor 2 \cdot s > e. \tag{7}$$

6.3.1.3. Heuristic Rules

Recall that the control rules gave us

$$\textbf{if } qq + ss \leq c \quad \textbf{then} \quad (q, qq) := (q + s, qq + ss)$$
$$\textbf{assert } s > e, \; qq \leq c$$
$$\textbf{else} \quad \textbf{assert } s > e, \; c < qq + ss \quad \textbf{fi}$$

but that the disjunction of $qq \leq c$ and $c < qq + ss$ turned out not to contribute anything. The heuristic

$<36>$ *conditional heuristic*
if t $\quad$ **then** $\quad P'$ $\qquad\qquad\qquad$ **assert** α $\qquad\quad$ **else** $\quad P''$ $\qquad\qquad\qquad$ **assert** β $\quad$ **fi** L:
L: **suggest** $\alpha, \; \beta$

suggests that each of the two invariants $qq \leq c$ and $c < qq + ss$ may be an invariant for both paths. So we have the candidates

> **suggest** $qq \leq c$, $c < qq + ss$

following the conditional and preceding the assignment

$$(s, ss) := (s/2, ss/2).$$

By application of the *forward assignment rule* $<21a>$ to the two candidates, we obtain

> **suggest** $qq \leq c$, $c < qq + 2 \cdot ss$

at the end of the loop.

Finally, by applying the *forward loop-body rule* $<29>$, we get the candidates

$$L_1^+ : \textbf{suggest} \quad \begin{aligned} &(q=0 \wedge qq=0 \wedge s=1 \wedge ss=d) \vee qq \leq c, \\ &(q=0 \wedge qq=0 \wedge s=1 \wedge ss=d) \vee c < qq + 2 \cdot ss \end{aligned}$$

Both candidates may be simplified, since their first disjunct implies their second, leaving

$$L_1^+ : \textbf{suggest} \quad qq \leq c, \ c < qq + 2 \cdot ss.$$

These two candidates can indeed be shown to be invariants. The first candidate, $qq \leq c$, derived from the initialization and **then-** paths, is unaffected by the **else**-path which leaves the value of qq unchanged. Similarly, the other candidate, $c < qq + 2 \cdot ss$, derived from the initialization and **else-** paths, is maintained true by the **then**-path. So we have the loop invariants

$$L_1^+ : \textbf{assert} \quad qq \leq c, \ c < qq + 2 \cdot ss. \tag{8}$$

Note that we have not yet made any use of the candidates

$$E_1^+ : \textbf{suggest} \quad q \leq c/d, \ c/d < q + e,$$

suggested by the output specification. For completeness, we shall apply a heuristic to these candidates, although no new invariants will be derived. The *top-down heuristic*

```
┌─────────────────────────────────┐
│ <38>  top-down heuristic        │
├─────────────────────────────────┤
│ suggest α                       │
│ loop L:                         │
│        until t                  │
│        P                        │
│        repeat                   │
│ suggest γ                       │
│                                 │
│ fact γ when α                   │
├─────────────────────────────────┤
│ L: suggest γ                    │
└─────────────────────────────────┘
```

suggests that the output candidate $q \leq c/d$ may itself be a loop invariant, since it is true upon entering the loop. Indeed it is, since it is implied by the loop invariant $qq \leq c$ and the global invariant $qq = q \cdot d$. On the other hand, the second output candidate, $c/d < q + e$, does not even hold for the initialization path, when $q = 0$.

Since there are no assignments between the loop and the end of the program, all the loop invariants may be pushed forward unchanged, and hold upon termination. The output invariants include

$$E_1^+ : \textbf{assert } s = 1 \lor 2 \cdot s > e, \ qq \leq c, \ c < qq + 2 \cdot ss, \ s \leq e. \qquad (9)$$

These invariants, along with the global invariants

$$\textbf{assert } ss = d \cdot s, \ qq = d \cdot q \ \textbf{in } T_1^+ ,$$

imply $q \leq c/d$ as specified. However, they do *not* imply $c/d < q + e$, only $c/d < q + 2 \cdot e$. In fact, our program as given is incorrect. In Chapter 3, we have already seen how such invariants are used to guide the debugging of a similar program.

6.3.1.4. Analysis

By introducing an "imaginary" loop counter n_1—initialized to 0 upon entering the loop and incremented by 1 with each iteration—one may derive relations between the program variables and the number of iterations. Such invariants may be useful for analyzing the efficiency of a program, as well as for proving correctness.[14]

The program T_1^+, annotated with some of the invariants we have already found and extended, is

---

assert $ss = d \cdot s$, $qq = d \cdot q$, $s \in 1/2^{N}$, $ss \in N/2^{N}$ **in**
T_1^{++} : **begin comment** *extended real-division program*
  B_1^{++} : **assert** $0 \le c < d$, $e > 0$
  $(q, qq, s, ss) := (0, 0, 1, d)$
  $n_1 := 0$
  loop L_1^{++} : **assert** $(s = 1 \lor 2 \cdot s > e)$, $qq \le c$, $c < qq + 2 \cdot ss$
    until $s \le e$
    if $qq + ss \le c$ **then** $(q, qq) := (q + s, qq + ss)$ **fi**
    $(s, ss) := (s/2, ss/2)$
    $n_1 := n_1 + 1$
    repeat
  E_1^{++} : **assert** $(s = 1 \lor 2 \cdot s > e)$, $qq \le c$, $c < qq + 2 \cdot ss$, $s \le e$
  end

Obviously, we may

$$\textbf{assert } n_1 \in N \textbf{ in } T_1^{++}.\tag{10}$$

For the variables s and n_1 we have the assignments

$$(s, n_1) := (1, 0) \mid (s, n_1) := (s/2, n_1 + 1)$$

and can apply the rule

<14> *linear rule*
$(x, y) := (a_0, b_0) \mid (x, y) := (x + a_2, b_1 \cdot y + b_2)$ **in** P
assert $[y \cdot (b_1 - 1) + b_2]^{a_2} \cdot b_1^{a_0} = [b_0 \cdot (b_1 - 1) + b_2]^{a_2} \cdot b_1^{x}$ **in** P

---

[14]Sometimes the analysis of a program is much more difficult than its verification; see, e.g. [JonassenKnuth78].

With this rule we get the global invariant

assert $[s \cdot (1/2 - 1) + 0]^1 \cdot (1/2)^0 = [1 \cdot (1/2 - 1) + 0]^1 \cdot (1/2)^{n_1}$ **in** T_1^{++} ,

which simplifies to

$$\text{\textbf{assert}} \ s = 1/2^{n_1} \ \textbf{in} \ T_1^{++} . \tag{11}$$

Applying the same rule to the assignments

$$(ss, n_1) := (d, 0) \ | \ (ss, n_1) := (ss/2, n_1 + 1)$$

we deduce

$$\text{\textbf{assert}} \ ss = d/2^{n_1} \ \textbf{in} \ T_1^{++} . \tag{12}$$

With these loop-counter invariants, the total number of loop iterations as a function of the input values may be determined. Using (11), we can substitute $1/2^{n_1}$ for s in the loop invariant (7) $s = 1 \vee 2 \cdot s > e$ and in the output invariant (9) $s \leq e$ and get $1/2^{n_1} = 1 \vee 2/2^{n_1} > e$ at L_1^{++} and $1/2^{n_1} \leq e$ at E_1^{++} . Taking the logarithm (e is positive), we have the upper bound

$$n_1 = 0 \vee n_1 < -\log_2 e + 1$$

and lower bound

$$-\log_2 e \leq n_1$$

on the number of loop iterations n_1. Note that by finding a loop invariant giving an upper bound on the number of iterations, we have proved that the loop terminates.

Combining both bounds at E_1^{++} gives (assuming $n_1 \neq 0$)

$$-\log_2 e \leq n_1 < -\log_2 e + 1,$$

or, since n_1 is an integer (10), it is equal to the one integer $\lceil -\log_2 e \rceil$ lying between its lower and upper bounds. Thus, we have the output invariant

$$E_1^{++} : \textbf{assert} \ n_1 = 0 \vee n_1 = \lceil -\log_2 e \rceil . \tag{13}$$

Since n_1 is the number of times the loop was executed before termination, we have derived the desired expression for the time complexity of the loop.

6.3.2. Array Minimum

In Section 5.3.1 we constructed—from specifications—a program to find the minimum value in an array segment, as well as its position. Here we begin with the same program and attempt to derive its output invariants from the code. The program was

$$
\begin{aligned}
&P_5^+: \textbf{begin comment } \textit{minimum program} \\
&\quad B_5^+: \textbf{assert } i \in Z,\ n \in Z,\ i \leq n \\
&\quad (k,j,m) := (n,n,A[n]) \\
&\quad \textbf{loop } L_5^+: \textbf{assert } \cdots \\
&\qquad \textbf{until } j{=}i \\
&\qquad j := j - 1 \\
&\qquad \textbf{if } A[j] < m \textbf{ then } (k,m) := (j,A[j]) \textbf{ fi} \\
&\qquad \textbf{repeat} \\
&\quad E_5^+: \textbf{suggest } A[k] \leq A[i{:}n],\ k \in [i{:}n],\ m{=}A[k] \\
&\textbf{end}
\end{aligned}
$$

It is supposed to find the position k and value m of the smallest element in the nonempty array segment $A[i{:}n]$.[15] We begin by applying assignment rules, followed by control rules.

6.3.2.1. Assignment Rules

We first try to determine the range of variables. The program variables are j, k, and m; the input variables i, n, and A are left unchanged.

The assignments to j in P_5^+ are

---

[15]Recall the abbreviation $A[k] \leq A[i{:}n]$ for $(\forall \xi \in [i{:}n]) A[k] \leq A[\xi]$.

$$j := n \mid j := j - 1.$$

Applying the *addition rule* $<1>$ to them, we obtain

$$\textbf{assert } j \in n - \mathbf{N} \textbf{ in } P_5^+$$

and consequently

$$\textbf{assert } j \in \mathbf{Z}, \ j \le n \textbf{ in } P_5^+ . \tag{1}$$

(since we are given that $n \in \mathbf{Z}$).

The assignments to k are

$$k := n \mid k := j.$$

Substituting the given range $\mathbf{N}$ for n and the derived (1) range $\mathbf{Z}$ for j, we can instead consider the nondeterministic assignments

$$k :\in \mathbf{N} \mid k :\in \mathbf{Z}$$

for k. Using the simple *set-union rule*

$<4>$ *set-union rule*
$x :\in S_0 \mid x :\in S_1 \textbf{ in } P$
$\textbf{assert } x \in S_0 \cup S_1 \textbf{ in } P$

it follows that k belongs to the union of n and $n - \mathbf{N}$, i.e. $k \in n - \mathbf{N}$. Thus, we may

$$\textbf{assert } k \in \mathbf{Z}, \ k \le n \textbf{ in } P_5^+ . \tag{2}$$

Finally, for m we have the assignments

$$m := A[n] \mid m := A[j].$$

Using $n \in \mathbf{Z}$ and (1) $j \in n - \mathbf{N}$ to substitute for i and j, we get

$$m :\in A[\mathbf{N}] \mid m :\in A[n - \mathbf{N}].$$

Thus, by the *set-union rule*, we obtain

$$\textbf{assert } m \in A[n - \mathbf{N}] \textbf{ in } P_5^+ . \tag{3}$$

If and when better bounds for j are derived, this invariant could be improved.

6.3.2.2. Control Rules

The program P_5^+ is an instance of the schema

> S^*: **begin comment** *single–loop schema*
>
> $z := c$
>
> **assert** $z=c$
>
> **loop** L^*: **assert** $z=c \lor s(g^-(z)) \lor {\sim}s(h^-(z))$,
> $\qquad\qquad z=c \lor {\sim}t(f^-(g^-(z))) \lor {\sim}t(f^-(h^-(z)))$
> $\qquad$ **until** $t(z)$
> $\qquad$ $z := f(z)$
> $\qquad$ **if** $s(z)$ **then** $z := g(z)$ **else** $z := h(z)$ **fi**
> $\qquad$ **assert** ${\sim}t(z_{L'})$, $[s(f(z_{L'})) \land z=g(f(z_{L'}))]$
> $\qquad\qquad\qquad\qquad\quad [{\sim}s(f(z_{L'})) \land z=h(f(z_{L'}))]$
> $\qquad$ **repeat**
> **end**

discussed in Section 6.2.4. In this case, z is the tuple (k,j,m), c is $(n,n,A[n])$, t is $(j=i)$, $f(k,j,m) = (k,j-1,m)$, $f^-(k,j,m) = (k,j+1,m)$, s is $(A[j]<m)$, $g(k,j,m) = (j,j,A[j])$, $g^-(k,j,m) = (\cdots,j,\cdots)$, and $h(k,j,m) = h^-(k,j,m) = (k,j,m)$. (The function g cannot be inverted to restore the values of k and m.) That gives the loop invariant

$$L_5^+ : \textbf{assert } (j=n \land k=n \land m=A[n]) \lor j+1\neq i \lor j+1\neq i \quad (4)$$

and, at the end of the loop body (after some simplifications):

$$\textbf{assert } (A[j]<m_{L_i^+} \land j=j_{L_i^+}-1 \land k=j \land m=A[j])$$
$$\lor (m \leq A[j] \land j=j_{L_i^+}-1 \land k=k_{L_i^+} \land m=m_{L_i^+}). \quad (5)$$

Applying the

> $<35>$ *forall rule*
>
> **assert** $x=a$
> **loop** L: **assert** $\alpha(x)$
> $\qquad\quad P$
> $\qquad\quad$ **assert** $x=f(x_L)$
> $\qquad\quad$ **repeat**
>
> L: **assert** $(\forall \xi \in \{a, f(a), \cdots, x\})\ \alpha(\xi)$

to (4), taking $\alpha(j)$ to be $j+1 \neq i$ and $f(j)$ to be $j-1$, gives $(\forall \xi \in [j:n])\, \xi+1 \neq i$, i.e. $i \leq j \lor i > n+1$, since i, j, and n are all integers. This yields

$$L_5^+ : \textbf{assert } j \in [i:n], \tag{6}$$

since we already know (1) that $j \leq n$ and it is given that $i \leq n$. And, since k is only assigned values of j we have, by the *set union rule* $<4>$,

$$L_5^+ : \textbf{assert } k \in [i:n], \tag{7}$$

as well.

If a relation α holds upon entering a loop, and the loop body either does not change the values of the variables in α, or reachieves α for the new values of the variables, then α is a loop invariant. This is expressed by the rule

$<34>$ *protected-invariant rule*
$\textbf{assert } \alpha(\overline{x})$ $\textbf{loop } L:$ $\quad P$ $\quad \textbf{assert } \alpha(\overline{x}) \lor \overline{x} = \overline{x}_L$ $\quad \textbf{repeat}$
$L: \textbf{assert } \alpha(\overline{x})$

By substituting k for j in the first disjunct of (5), we may derive $m = A[k]$. Thus, at the end of the loop body we know $m = A[k] \lor (k = k_{L_5^+} \land m = m_{L_5^+})$, which is of the form $\alpha(\overline{x}) \lor \overline{x} = \overline{x}_L$, taking $\alpha(k,m)$ to be $m = A[k]$. The first disjunct indicates that the **then**-path achieves α; the second disjunct states that the **else**-path leaves both k and m unchanged. Since α holds initially as well, we may

$$L_5^+ : \textbf{assert } m = A[k]. \tag{8}$$

Finally, we use the *forward loop-body rule* $<29>$ to determine that

$$L_5^+ : \textbf{assert } m \leq A[j], \tag{9}$$

regardless of which path was taken. This invariant holds for the initialization path, when $j = n$ and $m = A[n]$; it holds for the loop-body path (5), either because $m = A[j]$ for the **then**-branch, or from the negation of the conditional.

6.3.2.3. Generalization Heuristic

The *generalization heuristic* is particularly valuable for loops involving arrays:

$$
\begin{array}{|l|}
\hline
<37b>\ \textit{generalization heuristic} \\
\hline
\textbf{assert } x = a,\ x \in \mathbf{Z} \\
\textbf{loop } L:\ \textbf{assert } \alpha(x,y) \\
\quad P \\
\quad \textbf{assert } x = x_L - 1 \\
\quad \textbf{repeat} \\
\hline
L:\ \textbf{suggest } (\forall\, \xi \in [x{:}a])\ \alpha(\xi,y) \\
\hline
\end{array}
$$

To apply this heuristic, consider the loop invariant (9) and let $\alpha(j,m)$ be $m \leq A[j]$. Initially j is n, and at the end of the loop body (5) $j = j_{L_5^+} - 1$, so, as an invariant candidate, we try

$$L_5^+ : \textbf{suggest } (\forall\, \xi \in [j{:}n])\ m \leq A[\xi],$$

which we abbreviate as $m \leq A[j{:}n]$. Checking the candidate for the **then**- and **else**- paths determines that it is in fact an invariant; thus, we have

$$L_5^+ : \textbf{assert } m \leq A[j{:}n]. \tag{10}$$

To summarize, we have for loop invariants

$$L_5^+ : \textbf{assert } j \in [i:n], \ k \in [i:n], \ m = A[k], \ m \leq A[j:n]. \tag{11}$$

Using the *forward loop-exit rule* $<31>$, the invariants at L_5^+ may be propagated past the exit test $j = i$, obtaining

$$E_5: \textbf{assert } j = i, \ k \in [i:n], \ m = A[k], \ m \leq A[j:n], \tag{12}$$

as desired.

6.3.2.4. Analysis

To determine the number of times the program loop is executed, we extend the program with a loop counter n_5. By the *addition-relation rule* $<11>$, one can easily determine

$$\textbf{assert } n_5 = n - j \textbf{ in } P_5^+ , \tag{13}$$

since j is initialized to n and decremented by 1. We know from (12) that $j = i$ when the loop is exited, and it follows that the loop is executed $n - i$ times:

$$E_5: \textbf{assert } n_5 = n - i. \tag{14}$$

6.3.3. Selection Sort

In this example, we annotate an array-manipulation program with nested loops. The program

```
P_9: begin comment selection-sort program
  B_9: assert n ∈ N
  i := 0
  loop L_9: assert  · · ·
        until i ≥ n
        assert m = A[k], j ∈ [i:n], k ∈ [i:n] in
        P_5^+ : begin
                suggest i ∈ Z, n ∈ Z, i ≤ n
                (k,j,m) := (n,n,A[n])
                loop    L_5^+ : assert m ≤ A[j:n]
                        until j = i
                        j := j - 1
                        if A[j] < m then (k,m) := (j,A[j]) fi
                        repeat
                assert m ≤ A[i:n], m = A[k], k ∈ [i:n]
                end
        (A[k],A[i],i) := (A[i],m,i+1)
        repeat
  E_9: suggest (∀ ξ ∈ [0:n-1]) A[ξ] ≤ A[ξ+1], {A[0:n]}={A_{B_9}[0:n]}
  end
```

is intended to sort the array $A[0:n]$ of $n+1$ elements $A[0]$, $A[1]$, ..., $A[n]$ in ascending sequence. Thus, its output specification can be expressed as

$$E_9: \text{ suggest } (\forall \xi \in [0:n-1]) A[\xi] \leq A[\xi+1], \{A[0:n]\}=\{A_{B_9}[0:n]\}$$

where $\{A[0:n]\}=\{A_{B_9}[0:n]\}$ means that the multiset (bag) of elements currently in the array segment $A[0:n]$ is equal to the multiset of elements contained in the array when control was initially at B_9, i.e. $A[0:n]$ is a permutation of $A_{B_9}[0:n]$.

In general, it is easier to annotate nested loops inside-out. The inner loop of this sorting program is identical with program P_5^+ of the previous example. We can, therefore, start this example with an already annotated inner loop. But we must verify that the input requirements

$$\text{suggest } i \in Z, n \in Z, i \leq n$$

of P_5^+ is satisfied.

We begin by applying assignment and control rules, followed by heuristics. Then we extend the program with counters and analyze its time complexity.

6.3.3.1. Algorithmic Rules

The assignments to i are

$$i := 0 \mid i := i+1,$$

which by the *addition rule* $<1>$ gives the global invariant

$$\textbf{assert } i \in \textbf{N in } P_9. \tag{1}$$

This satisfies the first part of the input specification of P_5^+ .

We can derive an invariant $i < n$ at the entry point to P_5^+ from the negation of the outer-loop exit test, using the loop axiom $<20>$. Since i is constant within P_5^+ , $i < n$ and the output invariants of P_5^+ hold just prior to the assignment

$$(A[k],A[i],i) := (A[i],m,i+1).$$

Propagating them forward, the invariants $i < n$ and $k \in [i{:}n]$ become $i \leq n$ and $k \in [i-1{:}n]$, respectively, since i is incremented (by the *forward assignment rule* $<21a>$). To propagate invariants over an array assignment, there is a special assignment rule:

$<23>$ *forward array-assignment rule*
assert $\alpha(A,z)$ $A[y] := f(A[y],z)$ $L:$
$L:$ **assert** $\alpha(assign(A,y,f^-(A[y],z)),z)$

where the array function $assign(A,y,u)$ yields A with u replacing $A[y]$, and $f^-(f(A[y],z),z)=A[y]$.[16] This rule states that after the assignment the invariant still holds for all the elements of A, save $A[y]$; it also holds

---

[16]The *assign* function was used by [McCarthy62]; array assignments are also discussed in [Gries77]; rules for pointer assignments are given in [LuckhamSuzuki79].

for old value of $A[y]$, $f^-(A[y],z)$. In our case, the invariant $m \leq A[i:n]$ still holds after assigning $A[i]$ to $A[k]$, since $m \leq A[i]$ held before; however, after the assignment to $A[i]$, we only know $m \leq A[i+1:n]$, and after incrementing i, we have $m \leq A[i:n]$. The assignment $A[i] := m$ generates the invariant $A[i]=m$ (by the *assignment axiom* $<18>$), which becomes $A[i-1]=m$ when i is incremented. We have therefore the following invariants at the end of the outer-loop body:

$$\textbf{assert } i \leq n, \ k \in [i-1:n], \ m \leq A[i:n], \ A[i-1]=m. \tag{2}$$

Clearly upon entering the outer loop (by $<18>$) we can

$$\textbf{assert } i=0.$$

Thus, by the *forward loop-body rule* $<29>$, we have the outer-loop invariant

$$L_9: \textbf{ assert } i=0 \lor (i \leq n \land k \in [i-1:n] \land m \leq A[i:n] \land A[i-1]=m)$$

with the corollary

$$L_9: \textbf{ assert } i=0 \lor A[i-1] \leq A[i:n], \ i \leq n. \tag{3}$$

If we use the *forward loop-exit rule* $<31>$ to push $i \leq n$ past the exit test $i \geq n$ and out of the loop, we get the output invariants $i \leq n$ and $i \geq n$ at E_9, i.e.

$$E_9: \textbf{ assert } i=n. \tag{4}$$

At the same time, we obtain $i < n$ at the head of the loop body, satisfying the input specification $i \leq n$ for P_5^+ . Since it is given that $n \in \mathbf{N}$ for P_9 and n is constant, all the input requirements of P_5^+ are satisfied.

6.3.3.2. Heuristics

We can apply the *generalization heuristic* $<37b>$ to the outer-loop invariant (3) for the counter i, taking $\alpha(i,A)$ to be $i=0 \lor A[i-1] \leq A[i:n]$. Since i is initially 0, this yields the candidate

$$L_9: \textbf{ suggest } (\forall \xi \in [0:i]) \ (\xi=0 \lor A[\xi-1] \leq A[\xi:n]).$$

This is equivalent to

L_9: **suggest** $(\forall \xi \in [0{:}i-1]) \, A[\xi] \le A[\xi+1{:}n]$

and states, in effect, that the array elements $A[0{:}i-1]$ are sorted and that they are all smaller than the array elements $A[i{:}n]$. Though the array A is modified along the inner-loop exit path by the assignment

$$(A[k], A[i], i) := (A[i], m, i+1),$$

using $k \in [i{:}n]$ and $m = A[k]$ invariance along that path can be shown.[17] So we have the outer-loop invariant

$$L_9: \textbf{assert } (\forall \xi \in [0{:}i-1]) \, A[\xi] \le A[\xi+1{:}n]. \qquad (5)$$

This may be pushed out of the loop to E_9, and with (4), implies the first conjunct of the output specification,

$$(\forall \xi \in [0{:}n-1]) \, A[\xi] \le A[\xi+1].$$

The *top-down heuristic* $<38>$ suggests that the output specification $\{A[0{:}n]\} = \{A_{B_9}[0{:}n]\}$, which is obviously true initially, is itself a candidate at L_9. Since the assignments to A have the effect of exchanging the values of $A[k]$ and $A[i]$, we also have the invariant

$$L_9: \textbf{assert } \{A[0{:}n]\} = \{A_{B_9}[0{:}n]\}. \qquad (6)$$

6.3.3.3. Analysis

We have already determined in Section 6.3.2.4 that the loop in P_5^+ is always executed $n_5 = n-i$ times. To determine the time complexity of the whole sorting program, we add another counter n_9 to sum the *total* number of inner-loop executions.

---

[17]Since $i \le k$, the assignment to $A[k]$ cannot destroy the order of $A[0{:}i-1]$; similarly the assignment to $A[i]$ has no effect. Since both i and k are in the range $[i{:}n]$, the candidate implies that $A[0{:}i-1] \le A[i]$ and $A[0{:}i-1] \le A[k]$. So assigning $A[i]$ to $A[k]$ does not affect the relation. Lastly, since m is equal to the previous value of $A[k]$, assigning that value to $A[i]$ also preserves the invariant. The effect of the assignment is to exchange the values of $A[k]$ and $A[i]$.

The extended program, annotated with some of the more important loop and output invariants, is

```
assert i ∈ N,  {A[0:n]}={A_{B_9^+}[0:n]} in
P_9^+ : begin comment extended selection–sort program
    B_9^+ : assert n ∈ N
    i := 0
    n_9 := 0
    loop L_9^+ : assert i ≤ n,  (∀ ξ ∈ [0:i−1]) A[ξ] ≤ A[ξ:n]
        until i ≥ n
        assert j ∈ [i:n], k ∈ [i:n], m=A[k] in
        P_5^+ : begin
            (k,j,m) := (n,n,A[n])
            loop L_5^+ : assert m ≤ A[j:n]
                until j=i
                j := j−1
                if A[j] < m then (k,m) := (j,A[j]) fi
                repeat
            end
        (A[k],A[i],i) := (A[i],m,i+1)
        n_9 := n_9+ n−i
        repeat
    E_9^+ : assert i=n,  (∀ ξ ∈ [0:n−1]) A[ξ] ≤ A[ξ:n]
    end
```

With each outer-loop iteration, i.e. each time i is incremented by 1, the inner loop is iterated $n-i$ times. Using the[18]

```
<15> geometric rule
(x,y) := (a_0,b_0) | (x,y) := (x+a_1,y+b_1+b_2·x) in P
assert 2·a_1·(y − b_0)=(x− a_0)·(2·b_1+ b_2·(x+ a_0− a_1)) in P
```

and taking x to be i (initially 0 and incremented by 1) and y to be n_9 (incremented by $n-i$), we derive the invariant

$$\text{assert } 2 \cdot n_9 = 2 \cdot n \cdot i - i^2 + i \text{ in } P_9^+. \tag{7}$$

The outer loop is executed n times, since i effectively counts the number of iterations, and we have already determined (4) that upon termination

---

[18]Recall the use in Section 5.3.2 of the inverse of this rule.

$i = n$. Thus, when the outer loop is exited, we have

$$E_9^+ : \textbf{assert } n_9 = n \cdot (n + 1)/2, \tag{8}$$

which is the total number of inner-loop executions.

6.4. Discussion

In a sense, annotating programs is "putting the cart before the horse" as the whole tenor of "structured programming" stresses developing invariants hand in hand with the code, and not *ex post facto*, as annotation implies. Nevertheless, the development of automatic tools for annotation is important for a number of reasons.

- The real world contains many undocumented, underdocumented, and misdocumented programs. Even annotated programs appearing in structured-programming textbooks have fallen prey to error! A system that could help in documenting such programs could be of utility.

- Ultimately, it is the responsibility of the programmer to ensure the correctness of his product. Even if he uses one of the current prototype verification systems, he is required to supply most, if not all, of the necessary invariant assertions, in addition to guiding the theorem prover. The goal of automatic program annotation is to relieve the programmer of some of this burden. Agreed, no present or foreseeable system will discover very subtle invariants, or those based on deep mathematical theorems, but such invariants are likely to be uppermost in the programmer's mind anyway. It is the "obvious" invariants that are annoying to have to formulate, and indeed are often forgotten. Fortunately, it is just these invariants that would be easy for an automatic annotation system to derive. Similarly, invariants needed to demonstrate the absence of runtime errors are usually quite simple; there has already been some success in providing verification systems with the capability of generating them.

- Annotation methods may be used to discover important properties of programs other than correctness, the investigation of which is generally outside the scope of the programmer's expertise. For example, one may wish to analyze the complexity of an algorithm or compare the efficiency of two correct programs. Even simple programs are sometimes very difficult to analyze; invariants may be needed to facilitate such an analysis. Similarly, as we have observed, it may be necessary to verify properties of a program in the course of attempted transformations.

- Annotation research attempts to formalize intuitions that lie behind well-designed programs; thus, it has implications for program synthesis. Some of the same rules used to generate invariants from programs may be inverted to generate programs from invariants.

Chapter 7

General Discussion

The last rung in the ladder of knowledge is the most abstract and subtle of all.

—Saadia ben Joseph al-Fayyumi Gaon (The book of beliefs and opinions)

But analogy may be a deceitful guide.

—Charles Darwin (The origin of species)

Though analogy is often misleading, it is the least misleading thing we have.

—Samuel Butler (Notebooks, music, pictures and books)

Programming is a complex human activity, requiring skill and expertise of its practitioners. The idea of designing an interactive programming system in which a human programmer is assisted by a semi-automatic verifier/debugger was first suggested in [Floyd71]; more recently, the need for assorted "intelligent" programming aids has been expressed in many quarters. In our view, it is important to incorporate—as an integral part of such programming environments—methods for transforming programs in ways that do not necessarily preserve their semantics, in addition to correctness-preserving transformations. Our purpose in the chapters that preceded was to contribute to that goal by describing a unified approach to formal program manipulation based upon invariant assertions. (For a survey of various logic-based approaches to the different aspects of programming, see [Manna-Waldinger78].) We believe that the kinds of evolutionary processes that we have formalized play an important role in programming.

Such a program development system might include the six units shown in Figure 3. At each stage of program construction, there is a partially written program, annotated with invariant assertions and other comments, explaining how it is meant to work. Three units participate in all phases of program construction:

- The *synthesizer* transforms the specification of a program into pieces of code and/or new subgoals satisfying those specifications. Schematically,

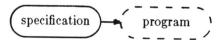

Even if the synthesizer produces only correct code, there can be no assurance that it will not embark on fruitless paths.

- The *modifier* has the task of transforming and augmenting existing programs to satisfy the specifications of a new program. Schematically,

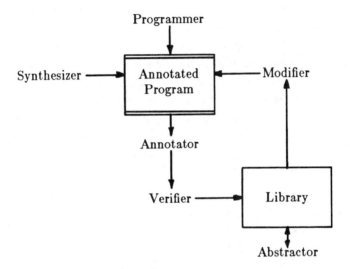

Figure 3. *Hypothetical semi-automated program-development system.*

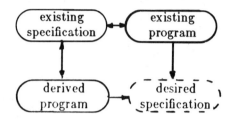

It may generate new subgoals if it finds that they are prerequisites for successful adaptation. Special cases of program modification are debugging, instantiation, extension, and optimization. For debugging, the source of error in a program must be located, and the incorrect program must be modified so as to correctly satisfy the original specifications. For instantiation, an applicable schema must be identified, and preconditions tested. For extension, a program that achieves some fraction of the specification is used as the basis for solving the complete problem. For optimization, a correct, but inefficient, program must be modified into a more efficient one, while preserving correctness.

- The *programmer,* whose insights and ingenuity may be beyond the capabilities of a fully automated system, lends his expertise to the task. On the other hand, by nature he is error-prone and his every move should therefore be closely monitored by a suspicious system.

An additional unit does not generate code, but rather analyzes given code and produces diagnostic information:

- The *verifier* constantly looks over the programmer's shoulder and checks for logical errors. If a section of code is found to be incorrect or nonterminating by the verifier, then it will need to be modified to do things right. Given an annotated program, the verifier proceeds to generate conditions for correctness and for incorrectness and attempts to prove one or the other. In addition, a program's efficiency ought to be estimated by determining time and space bounds on its performance.

It has been argued ([DeMilloLiptonPerlis79]) that formal verification is not what will make for reliable software. (For the pro side, see for example, [London77].) Published, "verified," and "tested" algorithms have been found to contain errors of all sorts ([GerhartYelowitz76a], [WeyukerOstrand79], and others). But a verifier is only one tool at a programmer's disposal; and no one tool is a panacea for all software problems. When an attempted verification shows that a program is inconsistent with its specification, then either the program or the specification is at fault. Only the programmer can know which. Furthermore, a well-designed verifier should be able to use its reasoning abilities to suggest possible locations for the error, not just answer "true or false." In our scenarios of modification and abstraction, we used unverifiable conditions to suggest additional transformations and preconditions.

A prerequisite for verification and modification is an appropriate set of invariant assertions. If the original program is not supplied with sufficient invariants to prove correctness or incorrectness, they must be generated from the program text, possibly using the specification and programmer-supplied assertions as a guide.

● The *annotator* generates invariants from the program text, describing what the program is actually doing and how. In a sense, this is the inverse of synthesis:

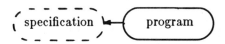

These invariants then enable the verifier to determine if what the program does is in fact what it should be doing. They also supply information about relationships between variables, needed to evaluate a program's efficiency.

Finally, one also needs the capability of extrapolating from programs written in the past. One would like to be able to extract programming techniques and strategies from given correct programs and previously constructed ones, and store them for future reference.

- The *abstractor* attempts to extract from finished program segments, the "essence" of the methods used. Abstraction goes a step further than modification, in that the specification is also derived from existing programs:

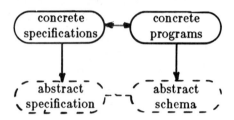

The abstractor may make use of program annotations and proofs to identify underlying techniques for abstraction, and to determine what are the necessary conditions for the schemata to be applicable. Derived schemata can then be instantiated by the modifier to solve future problems. In a similar manner, abstraction can be used to devise schematic program transformations, given concrete examples.

The examples of the previous chapters have illustrated how such a hypothetical automated-programming system might perform some of the varied tasks of program construction. The necessary reasoning abilities needed for each of the tasks are alike; the requisite manipulation abilities are similar to what would be required of any program-transformation system. Also, the different units all require knowledge of the subject domains. Unfortunately, all these tasks demand sophisticated theorem-proving capabilities, even for simple programs, capabilities that today's provers often lack. Invariants played a basic role throughout; in determining which transformations to perform, the program's invariants were compared with the desired invariants; they served as the basis for abstraction; during program synthesis, invariants were developed hand in hand with the code.

Program verification, synthesis, and annotation are intimately related; all three share a need for a similar understanding of programming methodology. When both the code and the assertions are given, then it is only necessary to prove that the assertions suffice to verify correctness. At the other extreme, for structured program-synthesis, only the specifications are given, and the code must be synthesized together with a proof. Program annotation lies midway between verification and synthesis in difficulty: given specifications, possible methods of achieving it need to be compared with the actual code. Thus, code and specifications serve to guide the search for invariants. We suspect that logic-based programming languages (see [Kowalski74]) may provide a relatively convenient test-bed for continued experiments in these and other forms of program manipulation. The identity of specification and programming languages, and the potential availability of suitable general-purpose theorem-provers, should aid the design of such systems.

By implementing some of our ideas, we hope to have demonstrated that they are in the realm of the feasible. With further work, we believe that the proposed methods can be made into practical programming tools. Some steps we took in our examples were more intuitively obvious than others, and more easily implementable. By themselves, the methods and strategies we have described cannot be expected to suffice for a fully automatic system to produce solutions to difficult programming problems. One major hurdle is in deciding which of many possible transformations to apply when and where. We do, however, envision the possibility of such methods being embedded in the kind of semi-automated program-development environment outlined above. The system would perform straightforward transformations and extensions in a consistent manner; the human would be left to guide the machine in the more creative steps.

Though the illustrations we have provided, and the examples we tested our system on, were small and limited in scope (our implementation quickly ran out of resources on anything larger), "top-down" programming methodology favors small, easily comprehensible modules, each of which should be amenable to the kind of manipulations we have presented. The larger a program, the more convoluted it is, or the deeper the ideas that went into it, the more difficult it would be—for man or machine—to reason about it. On the other hand, the detection and correction of various "runtime" errors, for example, requires substantially less reasoning ability than "full-fledged" annotation and debugging. Similarly, small segments of programs are more readily synthesizable than complete programs.

In this book, we have touched upon some of the evolutionary aspects of program construction. In particular, we have shown how analogies can be exploited to allow a system to profit from experience. But analogical reasoning is only one mode of thought—albeit an important one; to be effective, it will need to be combined with many other techniques.

Appendix 1

Global Transformations

He found himself transformed in his bed into a gigantic insect.

—*Franz Kafke (Metamorphosis)*

In this appendix, we present some rules for global program transformations. First, some definitions:

- A *constant* is any symbol appearing in a program with assumed properties, e.g. 0, **true**, +, and $\geq$.

- A *variable* is any symbol appearing in the program with no assumed properties other than those mentioned in the input specification.

- A *variable expression* is an expression containing at least one variable.

- A *constant expression* contains no variables.

- A variable that changes value during program execution is termed a *program variable*.

- Any other variable is considered to be an *input variable*.

- A program variable appearing in the output specification is called an *output variable*.

The transformations we use may be divided into several categories. Those that transform constants are not guaranteed to preserve correctness; to prove the validity of such transformations, the verification conditions of the transformed program must be shown to hold. On the other hand, transformations that systematically replace variables by variable expressions (of the appropriate type) are guaranteed to yield a program satisfying the transformed specifications. The validity of these transformations may be shown by proving that for a transformation $\mu \Rightarrow \nu$, the expression μ of the original program is invariantly equal to the expression ν of the new program. In other words, if the original program and the transformed program were run in parallel, after each statement executed, μ of the original program would equal ν at the corresponding point in the execution of the new program. In all other respects the program is unchanged. These transformations are not, however, guaranteed to result in a "clean" program, in the sense that the transformations may introduce nonprimitive operators or result in runtime errors such as taking the square-root of a negative number. For some transformations of the form $y \Rightarrow \nu(y)$, where ν is a function of y, the existence of an inverse function ν^- is necessary.

Let P be a program, and P' the same program after applying a transformation. Some correctness-preserving transformations are:

1) $a \Rightarrow \nu(\bar{b})$, where a is an input variable and $\nu(\bar{b})$ is a variable expression containing only input variables $\bar{b}$. All occurrences of a in the annotated program P are replaced by $\nu(\bar{b})$ in P'.

To see that the transformed program P' is correct, consider any legal input values $\bar{b}_0$. Assume, then, that program P is run with its input variable a equal to $\nu(\bar{b}_0)$. Since each occurrence of a in the input specification has been replaced by v, if v is a legal input for P', then a is a legal input for P. Since all occurrences of a in P have been replaced by v in P', any expression $e(a)$ in P has the same value as the transformed expression $e(\nu(\bar{b}_0))$ in P'. The execution of P' therefore parallels the execution of P. Thus, if the output specification of P holds for a, then the transformed specification holds for $\nu(\overline{b}_0)$.

2) $y \Rightarrow \nu(y)$, where y is a variable and $\nu(y)$ is an expression containing y (a program variable is replaced by an expression containing that variable). To apply this transformation, all occurrences of the variable y in the annotated program are replaced by $\nu(y)$. Where y appears on the left-hand side of an assignment $y := \epsilon$, the assignment is replaced by $y := \nu^-(\epsilon)$, where $\nu(\nu^-(u))=u$ for all u.

Let y' refer to the variable y in the transformed program P', as opposed to the variable y of the original program P. We prove that $y=\nu(y')$ is invariant by induction on the execution. When y is assigned to for the first time, the assignment $y := \epsilon$ in the original P is paralleled by the assignment $y' := \nu^-(\epsilon)$ in P'. This has the effect of giving $\nu(y')$ the value $\nu(\nu^-(\epsilon))=\epsilon$ which is the same value assigned to y; thus, $\nu(y')=y$, as is desired. Assume, then, that $y=\nu(y')$. Wherever the variable y occurs in an expression $\epsilon(y)$ in P, that expression has been replaced by $\epsilon(\nu(y'))$, which, by induction, has the same value $\epsilon(y)$ in P'. Therefore, all tests in P' have the same result as their counterparts in P. Furthermore, for any subsequent assignment to y, the assignment $y := \epsilon$ in P has the same effect on y as the assignment $y' := \nu^-(\epsilon)$ in P' has on v, since e has the same value in both programs.

3) $y \Rightarrow \nu(y,z)$, where y and z are program variables. If the variables y and z are not always assigned to in parallel, then add a trivial assignment $y := y$ or $z := z$. An assignment $(y,z) := (\delta,\epsilon)$ in P is then replaced by $(y,z) := (\nu^-(\delta,\epsilon),\epsilon)$ in P', where $\nu(\nu^-(u,v),v)=u$ for all u and v.

We prove that $y=\nu(y',z)$ is invariant. The proof is essentially the same as the previous one. When an assignment $(y,z) := (\delta,\epsilon)$ is executed in P, the assignment $(y',z) := (\nu^-(\delta,\epsilon),\epsilon)$ in P' has the effect of letting $\nu(y',z)=\nu(y',\epsilon)$ become $\nu(\nu^-(\delta,\epsilon),\epsilon)=\delta$, which is the new value of y.

4) $\mu(y) \Rightarrow \nu(y)$, where y is any variable. This is not in general a correctness-preserving transformation. However, if all the occurrences of the variable y (except those on the left-hand side of assignments), are within occurrences of the subexpression $\mu(y)$, then

correctness is preserved. When y appears on the left of an assignment $y := \epsilon$, that statement is replaced by $y' := \nu^-(\mu(\epsilon))$. In that case, it is easy to see that $\mu(y)=\nu(y')$.

There are several ways in which an expression of the form $f(q,r)$ can be compared with another expression h. If h is of the form $g(s,t)$, then the *imitating* mapping $f \Rightarrow g$, $q \Rightarrow s$, and $r \Rightarrow t$ suggests itself. Mapping $f \Rightarrow g$, however, does not necessarily preserve correctness and must be verified. If f has an inverse f^- in its first argument, i.e. $f(f^-(x,y),y)=x$ for all x and y, then the *inverting* mapping $q \Rightarrow f^-(h,r)$ would work. If f has an identity element f^0 in its first argument, i.e. $f(f^0,v)=v$, then the *collapsing* mapping $q \Rightarrow f^0$ and $r \Rightarrow h$ is possible. Another possibility is the *projecting* mapping $f \Rightarrow \pi$ and $r \Rightarrow h$, where π projects its first argument, i.e. $\pi(u,v)=u$. (Similar mappings work for other than the first argument.)

5) $y \Rightarrow z$, where y and z are distinct variable symbols. This is permissible if z is a new variable, not originally occurring in the program, since, in that case, y in P has the same value as z in P'. It is also permissible when z is not new, if such a substitution will not result in a contradictory assignment, i.e. if for all assignments $(y,z) := (\delta(y,z),\epsilon(y,z))$, it is the case that $\delta(z,z)=\epsilon(z,z)$. Then, the multiple assignment may be replaced by $z := \delta(z,z)$. Since $\delta(u,u)=\epsilon(u,u)$ for all assignments, by induction $y=z$.

6) $\mu(y) \Rightarrow z$, where y and z are any two distinct variables. This is just a combination of the previous two cases. It is permissible only if z is a new variable not appearing elsewhere in the program, or only at points where $\mu(y)=z$. An assignment $y := \epsilon$ may be replaced by $(y,z) := (\epsilon,\mu(\epsilon))$; if $\mu(y)$ occurs in $\mu(\epsilon)$, it too may be replaced by z.

Appendix 2

Program Schemata

*There is also an abstraction involved in naming an operation
and using it on account of "what it does" while completely
disregarding "how it works." There is a strong analogy
between using a named operation in a program regardless of
"how it works" and using a theorem regardless of how it has
been proved. Even if its proof is highly intricate, it may be a
very convenient theorem to use!*

—Edsger Wybe Dijkstra (Notes on structured programming)

In [Gerhart75b] the hand-compilation of a handbook of program
schemata is recommended. Such a collection of schemata, together with
a library of program transformations (see, for example [Standish, *et al.*76]),
could serve as part of an interactive program-development system. This
appendix contains fifteen representative schemata, culled from the programming literature.

We use the following nomenclature:

j,k,x,I,T	*input variables*
X	*set of legal inputs*
p,q,t	*predicate symbols (assumed to be total)*
c,d,e,f,g,h	*function symbols (assumed total)*
a,b	*constant symbols*
z,r	*output variables*
A	*array variable*
i,s,y,m,S,Y	*program variables*
u,v,w,U,V,W	*universally quantified variables*
	(quantifier usually omitted)
n	*existentially quantified variable*
R	*set of real numbers*
N	*set of natural numbers*
Z	*set of integers*
$[u:v]$	*set of integers between u and v*
$|U|$	*size of set U*

Each of the following schemata is followed by an output assertion giving its abstract input-output specification. They are preceded by an input assertion containing the preconditions for correct application. The general format is

```
Sᵢ: begin
    comment  title (source)
             purpose
    assert   type conditions,
             correctness preconditions,
             termination precondition
    . . .
    schema body
    . . .
    assert   output specification
    end
```

The references are to sources that present an abstract schema; they are somewhat arbitrary, as the methods themselves are all well known. (The schemata may differ in details from those given in the referenced sources.)

S_1: **begin**

   comment   *binary search* [DershowitzManna75]

                  use a binary search to

                  approximate transition from q true to q false

   assert $j,k,x \in \mathbf{R}$,

           $q(j),\ \sim q(k)$,

           $q(u+x) \wedge x \geq v \supset q(u+v)$,

           $x > 0$

   $(z,s) := (j,k-j)$

   loop assert $q(z),\ \sim q(z+s),\ j \leq z \leq z+s \leq k \vee j > k$

        until $s \leq x$

        $s := s/2$

        if $q(z+s)$ **then** $z := z+s$ **fi**

        repeat

   assert $q(z),\ \sim q(z+x),\ j \leq z \leq k \vee j > k$

   end

S_2: **begin**

   comment   *element by element* [Gerhart75b]

                  achieve q for all integers between j and k

                  by applying h to each element

   assert $j \in \mathbf{Z},\ k \in \mathbf{R},\ x \in X$,

        $h : [j{:}k] \times X \to X$,

        $u \in [j{:}k] \wedge v \in X \supset q(h(u,v),u)$,

         $u,w \in [j{:}k] \wedge v \in X \wedge u < w \wedge q(v,u) \supset q(h(w,v),u)$

   $(z,i) := (x,j)$

   loop assert $(\forall u \in [j{:}i{-}1])\ q(z,u),\ i \in [j{:}k{+}1] \vee i{=}j > k{+}1,\ z \in X$

        until $i > k$

        $(z,i) := (h(i,z),i{+}1)$

        repeat

   assert $(\forall u \in [j{:}k])\ q(z,u),\ z \in X$

   end

```
S₃: begin
    comment  gradient search [Misra78]
             search for local extremum of q
             with locality defined by T
    assert  x ∈ X,
            T : X → 2^X,
            u ∈ X ⊃ q(u,u),
            u,v,w ∈ X ∧ q(w,u) ∧ ~q(w,v) ⊃ q(v,u),
            u,v,w ∈ X ∧ q(u,w) ⊃ q(u,v) ∨ q(v,w),
            |X| < ∞
    comment  y :∈ S chooses an arbitrary element of S
    z := x
    loop (m,S) := (z,T(z))
         loop   assert (∀ u ∈ T(m)-S) q(z,u),
                       z,m ∈ X, S ⊆ X
                until S={}
                y :∈ S
                S := S -{y}
                if ~q(z,y) then z := y fi
                repeat
         assert (∀ u ∈ T(m)) q(z,u), z ∈ X
         until z=m
         repeat
    assert (∀ u ∈ T(z)) q(z,u), z ∈ X
    end
```

```
S₄: begin
    comment  linear search [Dijkstra72]
             find least integer not less than j satisfying t
    assert  j ∈ Z, (∃ n ∈ [j:∞]) q(n)
    z := j
    loop assert (∀ u ∈ [j:z-1]) ~t(u), z ∈ Z
         until t(z)
         z := z+1
         repeat
    assert z = min   t(n)
             n ∈ [j:∞]
    end
```

S_5: **begin**
   comment   *extremum* [DershowitzManna75]
              achieve q for all integers between j and k
              using transitive property of q
   assert $j \in \mathbf{Z},\ k \in \mathbf{R},$
         $u \in [j{:}k] \supset q(u,u),$
         $u,v,w \in [j{:}k] \wedge u < v \wedge q(w,u) \wedge {\sim}q(w,v) \supset q(v,u)$
$(z,i) := (j,j)$
loop **assert** $(\forall u \in [j{:}i])\ q(z,u),\ z,i \in [j{:}k] \vee i{=}j > k$
      until $i > k{-}1$
      $i := i{+}1$
      if ${\sim}q(z,i)$ **then** $z := i$ **fi**
      repeat
assert $(\forall u \in [j{:}k])\ q(z,u),\ z \in [j{:}k] \vee j > k$
end

S_6: **begin**
   comment   *invertible recursion* [Cooper66]
              compute recursive function $f(x)$
              using inverse g of d
   assert $f(a){=}c(a),$
         $u \ne a \supset f(u){=}h(f(d(u)),u),$
         $d(g(u)){=}u,$
         $(\exists n \in \mathbf{N})\ g^n(a){=}x$
$(z,y) := (c(a),a)$
loop **assert** $z{=}f(y),\ (\exists n \in \mathbf{N})\ g^n(y){=}x,$
                     $(\exists n \in \mathbf{N})\ g^n(a){=}y$
      until $y{=}x$
      $(z,y) := (h(z,g(y)),g(y))$
      repeat
assert $z{=}f(x)$
end

S_7: **begin**
  comment *associative recursion* [DarlingtonBurstall76]
               compute recursive function $f(x)$
               defined by associative function h
  assert $p(u) \supset f(u)=c(u),$
         $\sim p(u) \supset f(u)=h(f(e(u)),d(u)),$
         $h(u,h(v,w))=h(h(u,v),w),$
         $(\exists n \in \mathbb{N})\, p(e^n(x))$
  if $p(x)$ **then** $z := c(x)$
           else $(z,s) := (d(x),e(x))$
                  loop **assert** $h(f(s),z)=f(x)$
                        until $p(s)$
                        $(z,s) := (h(d(s),z),e(s))$
                        repeat
                  $z := h(c(s),z)$ **fi**
  assert $z=f(x)$
  end

S_8: **begin**
  comment *tail recursion* [McCarthy62]
               compute recursive function $f(x)$
               iteratively
  assert $p(u) \supset f(u)=c(u),$
         $\sim p(u) \supset f(u)=f(e(u)),$
         $(\exists n \in \mathbb{N})\, p(e^n(x))$
  $y := x$
  loop assert $f(y)=f(x)$
        until $p(y)$
        $y := e(y)$
        repeat
  $z := c(y)$
  assert $z=f(x)$
  end

S_9: **begin procedure** $sort(A[j:k])$
    comment   *sorting* [Darlington78]
                  sort array segment $A[j:k]$ recursively,
                  using f to partition or g to merge
    assert $A[j:k] \in X[j:k]$,
           $f : X[u:w] \rightarrow X[u:w] \times [u:w-1]$,
           $g : X[u:v] \times X[v+1:w] \rightarrow X[u:w]$,
           $U \in X[u:v] \wedge V \in X[v+1:w] \supset \{g(U,V)\} = \{U\} \cup \{V\}$,
           $U \in X[u:w] \wedge f(U) = (W,v) \supset \{U\} = \{W\}$,
           $U \in X[u:w] \wedge f(U) = (W[u:w],v)$
                        $\wedge \ sorted(W[u:v]) \wedge \ sorted(W[v+1:w])$
                            $\supset sorted(g(W[u:v], W[v+1:w]))$
    comment $X[u:v]$ is the set of arrays of elements of X
                  with indices in $[u:v]$
    comment $\{U\} = \{V\}$ means that for each occurrence
                  of an element in U there is an occurrence
                  of that element in V
    comment $sorted(W[u:w])$ means $(\forall v \in [u:w-1]) \ W[v] \le W[v+1]$
    comment A_B denotes the value of A upon procedure entry
    if $j < k$ **then** $(A[j:k],m) := f(A[j:k])$
                          $sort(A[j:m])$
                          $sort(A[m+1:k])$
                          $A[j:k] := g(A[j:m],A[m+1:k])$ **fi**
    assert $\{A_B[j:k]\} = \{A[j:k]\}$, $sorted(A[j:k])$
    end

S_{10}: **begin**
  comment *right–permutative recursion* [Cooper66]
             compute recursive function $f(x)$
             defined by right-permutative function h
  assert $p(u) \supset f(u) = a,$
        $\sim p(u) \supset f(u) = h(f(e(u)), d(u)),$
        $h(g(u,v), w) = g(h(u,w), v),$
        $h(a,u) = g(a,u),$
        $(\exists n \in \mathbf{N}) \, p(e^n(x))$
  $(z,y) := (a,x)$
  loop **assert** $f(x) = g(g(\;\cdots\;g(g(z, d(y)), d(e(y))), \cdots\;),$
$$d(e^{\min_{n \in \mathbf{N}} p(e^n(y)) - 1}(y)))$$
      until $p(y)$
      $(z,y) := (g(z, d(y)), e(y))$
      repeat
  assert $z = f(x)$
  end

S_{11}: **begin**
  comment *additive relation* [DershowitzManna81]
             achieve $p(z)$ maintaining associative-commutative relation
             g between inputs x and t and outputs z and r
  assert $g(u,v) = g(v,u),$
        $g(g(u,v), w) = g(u, g(v,w)),$
        $h(h(u,v), w) = h(h(u,w), v),$
        $h(g(u,v), w) = g(h(u,w), h(v,w)),$
        $(\exists n \in \mathbf{N}) \, p(d^n(x)),$ where $d(u) = g(u, h(f(u), a))$
  $(z,r) := (x,t)$
  loop **assert** $g(h(x,b), h(r,a)) = g(h(t,a), h(z,b))$
      until $p(z)$
      $(z,r) := (g(z, h(f(z), a)), g(r, h(f(z), b)))$
      repeat
  assert $p(z),$ $g(h(x,b), h(r,a)) = g(h(t,a), h(z,b))$
  end

```
S₁₂: begin
   comment  double recursion [Knuth74]
              compute recursive function f(x)
              defined in terms of an associative function h
              applied to two recursive calls to f
              using a stack
   assert  p(u) ⊃ f(u)=c(u),
           ~p(u) ⊃ f(u)=h(f(e(u)),f(d(u))),
           h(a,u)=u,
           h(u,h(v,w))=h(h(u,v),w),
           p(x) ∨ (∃ n ∈ N)gⁿ({x})={}, where g(U)=
              {e(v):v ∈ U ∧ ~p(v)}∪{d(v):v ∈ U ∧ ~p(v)}
   comment  if s is a list ⟨sₙ,sₙ₋₁, · · · ,s₁⟩:
              head(s)=sₙ; tail(s)=⟨sₙ₋₁, · · · ,s₁⟩;
              u○s=⟨u,sₙ, · · · ,s₁⟩; |s|=n
   (z,s) := (a,⟨x⟩)
   loop assert f(x)=h(z,h(s₁,h(s₂,h( · · · ,h(s|ₛ|₋₁,s|ₛ|) · · · ))))
        until s=⟨⟩
        (y,s) := (head(s),tail(s))
        if p(y)  then    z := h(z,c(y))
                 else    s := e(y)○d(y)○s        fi
        repeat
   assert z=f(x)
   end
```

S_{13}: **begin**

  comment *linear recursion* [Wossner,*etal.*78]

               compute recursive function $f(x)$

               using a stack

               (for a stackless schema see [PatersonHewitt70])

  assert $p(u) \supset f(u) = c(u),$

          $\sim p(u) \supset f(u) = h(f(d(u)),u),$

          $(\exists n \in \mathbf{N})\, p(d^n(x))$

  comment if s is a list $\langle s_n, s_{n-1}, \cdots, s_1 \rangle$:

           $head(s) = s_n;\ tail(s) = \langle s_{n-1}, \cdots, s_1 \rangle;$

           $u \circ s = \langle u, s_n, \cdots, s_1 \rangle;\ |s| = n$

  $(y,s) := (x, \langle \rangle)$

  loop assert $f(x) = h(\cdots h(h(f(y), s_{|s|}), s_{|s|-1}) \cdots, s_1)$

      until $p(y)$

      $(y,s) := (d(y), y \circ s)$

      repeat

  $z := c(y)$

  loop assert $f(x) = h(\cdots h(h(z, s_{|s|}), s_{|s|-1}) \cdots, s_1)$

      until $s = \langle \rangle$

      $(z,s) := (h(z, head(s)), tail(s))$

      repeat

  assert $z = f(x)$

  end

S_{14}: **begin**
  comment *marking* [YelowitzDuncan77]
          collect elements related by T to I
  assert $I \subseteq X$, $T \subseteq X \times X$,
          $f : X \times 2^X \times 2^X \rightarrow 2^X$,
          $\{n : (u,n) \in W\} \subseteq f(u,V,W) \subseteq V$,
          $|T^*(I)| < \infty$
  comment $T^*(U)$ is the image of U under the reflexive-transitive
          closure of T, i.e.
          $T^*(U) = U \cup T(U) \cup T^2(U) \cup \cdots$,
          where $T(U) = \underset{u \in U}{\cup} \{n : (u,n) \in T\}$
  comment $y :\in S$ chooses an arbitrary element of S
  $(Z,S,Y) := (I,I,T)$
  loop **assert** $S \subseteq Z \subseteq T^*(I) \subseteq Z \cup Y^*(S) \subseteq X$, $Y \subseteq T$
        until $Y=\{\} \vee S=\{\}$
        $m :\in S$
        $S := S \cup f(m,Z,Y) - \{m\}$
        $Z := Z \cup \{u : (m,u) \in Y\}$
        $Y := Y - \{(u,v) : (m,v) \in Y\}$
        repeat
  assert $Z = T^*(I)$
  end

S_{15}: **begin procedure** *back* (T,x,Z)
  comment *backtracking* [GerhartYelowitz76b]
        collect elements related to x by T that satisfy p
  assert $x \in X$,
        $T : X \to 2^X$,
        $|T^*(\{x\})| < \infty$
  comment $T^*(U)$ is the image of U under the reflexive-transitive
        closure of T, i.e.
        $T^*(U) = U \cup T(U) \cup T^2(U) \cup \cdots$,
          where $T(U) = \bigcup_{u \in U} T(u)$
  comment Z_B denotes the value of Z upon procedure entry
  comment $y :\in S$ chooses an arbitrary element of S
  if $p(x)$ **then** $Z := Z \cup \{x\}$ **fi**
  $Y := T(x)$
  loop assert $Z_B \cup \{u \in T^*(\{x\}): p(u)\} = Z \cup \{u \in T^*(Y): p(u)\}$
     until $Y = \{\}$
     $m :\in Y$
     $Y := Y - \{m\}$
     back (T,m,Z)
     repeat
  assert $Z = Z_B \cup \{u \in T^*(\{x\}): p(u)\}$
  end

Appendix 3

Synthesis Rules

No rule so good as the rule of thumb, if it fit.

—Scottish proverb

This appendix contains top-down rules for synthesizing programs from specifications. Each rule is of the form

$<$x$>$ *name*
achieve $\alpha(\overline{u},\overline{v})$ **varying** $\overline{v}$
purpose $\alpha(\overline{u},\overline{v})$ *code containing new subgoals*   **assert** $\alpha(\overline{u},\overline{v})$

and applies to a subgoal of the form

$$\textbf{achieve } \alpha(\overline{u},\overline{v}) \textbf{ varying } \overline{v},$$

where α is a relation that should be made to hold among the variables $\overline{u}$ and $\overline{v}$ but only by setting the values of $\overline{v}$. The rule replaces the chosen subgoal with a segment of code—that may itself contain additional **achieve** statements. The new code has as its purpose

$$\textbf{purpose } \alpha(\overline{u},\overline{v});$$

if all the **achieve** statements in the new code are resolved, then the desired relation

$$\textbf{assert } \alpha(\overline{u},\overline{v})$$

will hold.

The eleven rules fall into the following categories:

- *To construct assignment statements:* $<a>$.

- *To construct conditional statements:* $<c>$.

- *To construct loop statements:* $<b>$, $<f>$, $<m>$, $<r>$, $<t>$.

- *To construct statement sequences:* $<d>$, $<p>$, $<v>$.

- *To transform goals:* $<s>$.

In rules $<m>$, $<r>$, and $<t>$, the ordering $>$ on W must be well-founded. In those rules, $\overline{v}_B$ refers to the values of the variables $\overline{v}$ upon entering the loop, $\overline{v}_E$ refers their values upon exiting the loop, and $\overline{v}'$ refers to the values prior to the current statement.

\<a\> *assignment rule*

---

achieve $y_1 = f_1(\overline{x}),\ y_2 = f_2(\overline{x}),\ \cdots,\ y_n = f_n(\overline{x})$
       varying $y_1, y_2,\ \cdots, y_n,\ \cdots$

---

purpose $y_1 = f_1(\overline{x}),\ y_2 = f_2(\overline{x}),\ \cdots,\ y_n = f_n(\overline{x})$
  $(y_1, y_2,\ \cdots, y_n) := (f_1(\overline{x}), f_2(\overline{x}),\ \cdots, f_n(\overline{x}))$
assert $y_1 = f_1(\overline{x}),\ y_2 = f_2(\overline{x}),\ \cdots,\ y_n = f_n(\overline{x})$

\<b\> *backward iterative loop rule*

---

achieve $\beta(\overline{u}, \overline{v})$ **preserving** $\alpha(\overline{u}, \overline{v})$ **for** $\overline{v}$
       varying $\overline{u}, \overline{v}$

---

purpose $\beta(\overline{u}, \overline{v})$
  loop $L:$ **purpose** $\alpha(\overline{u}, \overline{v})$ **for** $\overline{v}$
           until $\beta(\overline{u}, \overline{v})$
           approach $\beta(\overline{u}, \overline{v})$ **preserving** $\alpha(\overline{u}, \overline{v})$ **for** $\overline{v}$
                     varying $\overline{u}, \overline{v}$

           repeat
assert $\beta(\overline{u}, \overline{v})$

\<c\> *conditional rule*

---

purpose α
  achieve β, γ **varying** $\overline{v}$

---

purpose α
  if β **then** **achieve** γ **protecting** β
                             varying $\overline{v}$
          else   **assert** $\sim\!\beta$
                   achieve α **varying** $\overline{v}$ **fi**
assert β, γ

\<d\> *disjoint goal rule*

---

achieve $\beta(\overline{v})$, $\gamma(\overline{u}, \overline{v})$ **varying** $\overline{u}, \overline{v}$

---

purpose $\beta(\overline{v})$, $\gamma(\overline{u}, \overline{v})$
  achieve $\beta(\overline{v})$ **varying** $\overline{v}$
  achieve $\gamma(\overline{u}, \overline{v})$ **varying** $\overline{u}$
assert $\beta(\overline{v})$, $\gamma(\overline{u}, \overline{v})$

$\langle f \rangle$ *forward iterative loop rule*

---

achieve β **protecting** α
                varying $\overline{v}$

---

purpose β, α
   loop L: **assert** α
          until β
          approach β **protecting** α
                          varying $\overline{v}$
          repeat
assert β, α

$\langle m \rangle$ *backward termination rule*

---

assert $\overline{v}_B \in W$, $\overline{v}_E \in W$, $\overline{v}_B \geq \overline{v}_E$
approach $\beta(\overline{v})$ **preserving** $\alpha(\overline{v})$ **for** $\overline{u}$
                   varying $\overline{v}$

---

assert $\sim\beta(\overline{v})$
achieve $\overline{v} < \overline{v}'$ **preserving** $\alpha(\overline{v})$, $\overline{v} \in W$ **for** $\overline{u}$
                   varying $\overline{v}$

$\langle p \rangle$ *protection rule*

---

achieve $\beta(\overline{v})$, $\gamma(\overline{u},\overline{v})$ **varying** $\overline{u},\overline{v}$

---

purpose $\beta(\overline{v})$, $\gamma(\overline{u},\overline{v})$
   achieve $\beta(\overline{v})$ **varying** $\overline{v}$
   achieve $\gamma(\overline{u},\overline{v})$ **protecting** $\beta(\overline{v})$
                          varying $\overline{u},\overline{v}$
assert $\beta(\overline{v})$, $\gamma(\overline{u},\overline{v})$

```
┌─────────────────────────────────────────┐
│ <r>  recursive loop rule                 │
├─────────────────────────────────────────┤
│ assert  γ(ū), ū ∈ W                      │
│ purpose  α(ū)                            │
│        . . .                             │
│     assert  γ(v̄), v̄ ∈ W, v̄ < ū          │
│     achieve  α(v̄) varying v̄              │
│        . . .                             │
│ assert  α(ū)                             │
├─────────────────────────────────────────┤
│ assert  γ(ū), ū ∈ W                      │
│ P(ū): procedure                          │
│     purpose  α(ū)                        │
│        . . .                             │
│        assert  γ(v̄), v̄ ∈ W, v̄ < ū       │
│        purpose  α(v̄)                     │
│            P(v̄)                          │
│        assert  α(v̄)                      │
│        . . .                             │
│     assert  α(ū)                         │
│     end                                  │
└─────────────────────────────────────────┘
```

```
┌─────────────────────────────────────────┐
│ <s>  strengthening rule                  │
├─────────────────────────────────────────┤
│ assert  γ                                │
│ achieve  α(ū) varying ū                  │
│                                          │
│ fact  α(ū) when  β(ū,v̄),  γ              │
├─────────────────────────────────────────┤
│ assert  γ                                │
│ purpose  α(ū)                            │
│     achieve  β(ū,v̄) varying ū,v̄          │
│ assert  α(ū)                             │
└─────────────────────────────────────────┘
```

```
┌─────────────────────────────────────────┐
│ <t>  forward termination rule            │
├─────────────────────────────────────────┤
│ assert  v̄_B ∈ W, v̄_E ∈ W, v̄_B ≥ v̄_E     │
│ approach  β(v̄)    protecting  α(v̄)       │
│                   varying  v̄             │
├─────────────────────────────────────────┤
│ assert  ∼β(v̄)                            │
│ achieve  v̄ < v̄′   protecting  α(v̄), v̄ ∈ W│
│                   varying  v̄             │
└─────────────────────────────────────────┘
```

$<\text{v}>$ *preservation rule*
purpose $\alpha(\overline{u})$ **achieve** $\beta(\overline{v})$, $\gamma(\overline{u},\overline{v})$ **varying** $\overline{u},\overline{v}$

purpose $\alpha(\overline{u})$
    achieve $\beta(\overline{v})$ **varying** $\overline{v}$
    achieve $\gamma(\overline{u},\overline{v})$ **preserving** $\gamma(\overline{u},\overline{v})$ **for** $\overline{v}$
                        varying $\overline{u},\overline{v}$
assert $\alpha(\overline{u})$

Appendix 4

Annotation Rules

> *The practical men believed that the idol whom they worship—rule of thumb—has been the source of the past prosperity, and will suffice for the future.*
>
> *—Thomas Henry Huxley*

In this appendix we present a catalog of annotation rules. This list is representative of the kinds of rules that may be used for annotation; it is not meant to be complete.

We use the following conventions:

- P, P', and P'' denote program segments;

- L, L', and L'' are statement labels;

- α, β, and γ denote predicates;

- x, y, and z are variables;

- a, a_i, and b_i are expressions which are constant in the given program segment;

- u and v are arbitrary expressions;

- f and g are arbitrary functions;

- $\bar{x}$, $\bar{y}$, $\bar{z}$, $\bar{u}$, and $\bar{v}$ may be vectors;

- **N** denotes the set of natural numbers and **Z** the set of all integers.

4.1. Assignment Rules

The assignment rules all yield global invariants. The notation

$$\bar{x} := \bar{u}_1 \mid \bar{x} := \bar{u}_2 \mid \cdots \text{ in } P$$

means that the listed assignments are all the assignments to elements of $\bar{x}$ within P.

4.1.1. Range Rules

$<1>$ *addition rule*
$x := a_0 \mid x := x + a_1 \mid x := x + a_2 \mid \cdots$ **in** P
assert $x \in a_0 + a_1 \cdot \mathbf{N} + a_2 \cdot \mathbf{N} + \cdots$ **in** P

$<2>$ *multiplication rule*
$x := a_0 \mid x := x \cdot a_1 \mid x := x \cdot a_2 \mid \cdots$ **in** P
assert $x \in a_0 \cdot a_1^{\mathbf{N}} \cdot a_2^{\mathbf{N}} \cdot \cdots$ **in** P

$<3>$ *exponentiation rule*
$x := a_0 \mid x := x^{a_2} \mid \cdots$ **in** P
assert $x \in a_0^{a_1^{\mathbf{N}} \cdot a_2^{\mathbf{N}} \cdots}$ **in** P

4.1.2. Set Assignment Rules

- $x :\in S$ assigns some u to x such that $u \in S$.

- ΣS is the closure of the set S under $+$.

- ΠS is the closure of the set S under $\cdot$.

$<4>$ *set-union rule*
$x :\in S_0 \mid x :\in S_1 \mid x :\in S_2 \mid \; \cdots \;$ **in** P
assert $x \in S_0 \cup S_1 \cup S_2 \cup \; \cdots \;$ **in** P

$<5>$ *set-addition rule*
$x :\in S_0 \mid x :\in x + S_1 \mid x :\in x + S_2 \mid \; \cdots \;$ **in** P
assert $x \in S_0 + \Sigma S_1 + \Sigma S_2 + \; \cdots \;$ **in** P

$<6>$ *set-multiplication rule*
$x :\in S_0 \mid x :\in x \cdot S_1 \mid x :\in S_2 \mid \; \cdots \;$ **in** P
assert $x \in S_0 \cdot \Pi S_1 \cdot \Pi S_2 \cdot \; \cdots \;$ **in** P

$<7>$ *set-exponentiation rule*
$x :\in S_0 \mid x :\in x^{S_1} \mid x :\in x^{S_2} \mid \; \cdots \;$ **in** P
assert $x \in S_0^{\Pi S_1 \cdot \Pi S_2 \cdot \cdots}$ **in** P

4.1.3. Counter Relation Rules.

● $f(y)$ and $g(y)$ are expressions containing the one variable y.

● $g^-(u)$ is the inverse of $g(u)$, i.e. $g^-(g(u))=u$.

$<8a>$ *addition-counter rule*

$(x,y) := (a_0,b_0) \mid (x,y) := (x+f(y),g(y))$ **in** P

assert $x=a_0+f(b_0)+f(g(b_0))+f(g(g(b_0)))+\cdots+f(g^-(y))$ **in** P

$<8b>$ *integer addition-counter rule*

$(x,y) := (a_0,b_0) \mid (x,y) := (x+f(y),y+1)$ **in** P
assert $y\in\mathbf{Z}$ **in** P

assert $x=a_0+\Sigma_{\xi=b_0}^{y-1}f(\xi)$ **in** P

$<9a>$ *multiplication-counter rule*

$(x,y) := (a_0,b_0) \mid (x,y) := (x\cdot f(y),g(y))$ **in** P

assert $x=a_0\cdot f(b_0)\cdot f(g(b_0))\cdot f(g(g(b_0)))\cdot\ \cdots\ \cdot f(g^-(y))$ **in** P

$<9b>$ *integer multiplication-counter rule*

$(x,y) := (a_0,b_0) \mid (x,y) := (x\cdot f(y),y+1)$ **in** P
assert $y\in\mathbf{Z}$ **in** P

assert $x=a_0\cdot\Pi_{\xi=b_0}^{y-1}f(\xi)$ **in** P

$<$10a$>$ *exponentiation-counter rule*

$(x,y) := (a_0,b_0) \mid (x,y) := (x^{f(y)}, g(y))$ **in** P

assert $x = a_0^{f(b_0) \cdot f(g(b_0)) \cdot f(g(g(b_0))) \cdots \cdots f(g^-(y))}$ **in** P

$<$10b$>$ *integer exponentiation-counter rule*

$(x,y) := (a_0,b_0) \mid (x,y) := (x^{f(y)}, y+1)$ **in** P
assert $y \in \mathbf{Z}$ **in** P

assert $x = a_0^{\Pi_{\xi=b_0}^{y-1} f(\xi)}$ **in** P

4.1.4. Basic Relation Rules

● The logarithms in consequents of relation rules can be to any base.

$<$11$>$ *addition-relation rule*

$(x,y) := (a_0,b_0)$
$\mid (x,y) := (x+a_1 \cdot u, y+b_1 \cdot u)$
$\mid (x,y) := (x+a_1 \cdot v, y+b_1 \cdot v) \mid \cdots$ **in** P

assert $a_1 \cdot (y-b_0) = b_1 \cdot (x-a_0)$ **in** P

$<$12$>$ *multiplication-relation rule*

$(x,y) := (a_0,b_0)$
$\mid (x,y) := (x \cdot u^{a_1}, y \cdot u^{b_1})$
$\mid (x,y) := (x \cdot v^{a_1}, y \cdot v^{b_1}) \mid \cdots$ **in** P

assert $x^{b_1} \cdot b_0^{a_1} = a_0^{b_1} \cdot y^{a_1}$ **in** P

<13> *exponentiation-relation rule*

$(x,y) := (a_0, b_0) \mid (x,y) := (x^{a_1^u}, y^{b_1^u})$

$\mid (x,y) := (x^{a_1^v}, y^{b_1^v}) \mid \cdots$ **in** P

assert $\log(x)^{\log(b_1)} \cdot \log(b_0)^{\log(a_1)} = \log(a_0)^{\log(b_1)} \cdot \log(y)^{\log(a_1)}$ **in** P

4.1.5. Assorted Relation Rules

<14a> *linear rule*

$(x,y) := (a_0, b_0) \mid (x,y) := (a_1 \cdot x + a_2, b_1 \cdot y + b_2 + b_3 \cdot x)$ **in** P

assert $[(y \cdot (b_1 - 1) + b_2) \cdot (a_1 - b_1) - b_3 \cdot (x \cdot (b_1 - 1) + a_2)]^{\log(a_1)}$
$\cdot [a_0 \cdot (a_1 - 1) + a_2]^{\log(b_1)}$
$= [(b_0 \cdot (b_1 - 1) + b_2) \cdot (a_1 - b_1) - b_3 \cdot (a_0 \cdot (b_1 - 1) + a_2)]^{\log(a_1)}$
$\cdot [x \cdot (a_1 - 1) + a_2]^{\log(b_1)}$ **in** P

---

provided $a_1 \neq 1$, $b_1 \neq 1$, and $a_1 \neq b_1$.

<14b> *parallel linear rule*

$(x,y) := (a_0, b_0) \mid (x,y) := (a_1 \cdot x + a_2, a_1 \cdot y + b_2 + b_3 \cdot x)$ **in** P

assert $(a_1 - 1) \cdot [(a_1 - 1) \cdot (y \cdot a_0 - x \cdot b_0) + a_2 \cdot (y - b_0)]$
$- (x - a_0) \cdot [b_2 \cdot (a_1 - 1) - a_2 \cdot b_3]$
$= (b_3 / a_1) \cdot [(a_1 - 1) \cdot a_0 + a_2] \cdot [(a_1 - 1) \cdot x + a_1]$
$\cdot [\log((a_1 - 1) \cdot x + a_2) - \log((a_1 - 1) \cdot a_0 + a_2)] / \log(a_1)$ **in** P

---

provided $a_1 \neq 1$.

$\langle 14c \rangle$ *simple linear rule*

$(x,y) := (a_0, b_0) \mid (x,y) := (x + a_2, b_1 \cdot y + b_2 + b_3 \cdot x) \textbf{ in } P$

assert $[y \cdot (b_1 - 1) + b_3 \cdot x + b_2 + b_3 \cdot a_2 / (b_1 - 1)]^{a_2 \cdot b_1^{a_0}}$
$$= [b_0 \cdot (b_1 - 1) + b_3 \cdot a_0 + b_2 + b_3 \cdot a_2 / (b_1 - 1)]^{a_2 \cdot b_1^{x}} \textbf{ in } P$$

---

provided $b_1 \neq 1$.

$\langle 14d \rangle$ *degenerate linear rule*

$(x,y) := (a_0, b_0) \mid (x,y) := (a_1 \cdot x + a_2, y + b_2 + b_3 \cdot x) \textbf{ in } P$

assert $a_1^{(y - b_0) \cdot (a_1 - 1)} \cdot [a_0 \cdot (a_1 - 1) + a_2]^{b_2 \cdot (a_1 - 1) - a_2 \cdot b_3}$
$$= a_1^{(x - a_0) \cdot b_3} \cdot [x \cdot (a_1 - 1) + a_2]^{b_2 \cdot (a_1 - 1) - a_2 \cdot b_3} \textbf{ in } P$$

$\langle 15 \rangle$ *geometric rule*

$(x,y) := (a_0, b_0) \mid (x,y) := (x + a_1, y + b_1 + b_2 \cdot x + b_3^{x}) \textbf{ in } P$

assert $2 \cdot a_1 \cdot (b_3^{x} - b_3^{a_0})$
$$= (b_3^{a_1} - 1) \cdot [2 \cdot a_1 \cdot (y - b_0) - (x - a_0) \cdot (2 \cdot b_1 + b_2 \cdot (x + a_0 - a_1))] \textbf{ in } P$$

<16> *factorial rule*

$(x,y) := (a_0 \cdot a_1, b_0) \mid (x,y) := (x + a_1, y \cdot x \cdot b_1)$ **in** P

---

assert $y = b_0 \cdot (a_1 \cdot b_1)^{x/a_1 - a_0} \cdot a_0^{x/a_1 - a_0}$ **in** P

---

where $u^{\bar{n}} = u \cdot (u+1) \cdot \cdots \cdot (u+n-1)$.

<17> *multiplication-exponentiation rule*

$(x,y) := (a_0, b_0)$
$\mid (x,y) := (x \cdot a_1^u, y^{b_1^u})$
$\mid (x,y) := (x \cdot a_1^v, y^{b_1^v}) \mid \cdots$ **in** P

---

assert $[x/a_0]^{\log(b_1)} = [\log(y)/\log(b_0)]^{\log(a_1)}$ **in** P

4.2. Control Rules

The control rules all yield local invariants. Control rules expressed with **assert** statements in the antecedent may also be applied to **suggest** statements; in that case, the consequent is only a candidate.

4.2.1. Control Axioms

<18> *assignment axiom*

$\bar{x} := \bar{a}$

---

assert $\bar{x} = \bar{a}$

---

where $\bar{a}$ is a constant expression, i.e. $\bar{x}$ do not appear within it.

<19> *conditional axiom*

if t **then** **assert** t
P'
else **assert** $\sim t$
P'' **fi**

<20> *loop axiom*

loop P'
until t
assert $\sim t$
P''
repeat
assert t

4.2.2. Assignment Control Rules

- A is an array variable.

- The array function $assign(A,y,z)$ yields A, with z replacing $A[y]$.

<21a> *forward assignment rule*

assert $\alpha(\overline{x},\overline{y})$
$\overline{x} := f(\overline{x},\overline{y})$
$L:$
$L:$ **assert** $\alpha(f^-(\overline{x},\overline{y}),\overline{y})$

where f^- is the inverse of the function f, i.e. $f^-(f(\overline{x},\overline{y}),\overline{y})=\overline{x}$.

<21b> *variant forward assignment rule*

assert $\alpha(\overline{u},\overline{y})$
$\overline{x} := \overline{u}$
$L:$
$L:$ **assert** $\alpha(\overline{x},\overline{y})$

where $\overline{x}$ do not appear in $\alpha(\overline{u},\overline{y})$.

$<21c>$ *simple forward assignment rule*
assert $\overline{x}=\overline{a}$ $\overline{x} := f(\overline{x},\overline{y})$ $L:$
$L:$ **assert** $\overline{x}=f(\overline{a},\overline{y})$

where $\overline{a}$ are constant expressions, i.e. $\overline{x}$ do not appear within them.

$<22>$ *backward assignment rule*
$L:$ $\overline{x} := \overline{u}$ **assert** $\beta(\overline{x},\overline{y})$
$L:$ **assert** $\beta(\overline{u},\overline{y})$

$<23>$ *forward array-assignment rule*
assert $\alpha(A,\overline{z})$ $A[y] := f(A[y],\overline{z})$ $L:$
$L:$ **assert** $\alpha(assign(A,y,f^-(A[y],\overline{z})),\overline{z})$

where $f^-(f(A[y],\overline{z}),\overline{z})=A[y]$.

$<24>$ *backward array-assignment rule*
$L:$ $A[y] := u$ **assert** $\beta(A,\overline{z})$
$L:$ **assert** $\beta(assign(A,y,u),\overline{z})$

4.2.3. Conditional Control Rules

$<25>$ *forward test rule*

assert α
if t **then** L':
                 P'
       else L'':
                 P'' **fi**

L': **assert** α, t L'': **assert** $\alpha, \sim t$

$<26>$ *backward test rule*

L:
if t **then**   **assert** α
                 P'
       else   **assert** β
                 P'' **fi**

L: **assert** $t \supset \alpha, \sim t \supset \beta$

$<27>$ *forward branch rule*

if t **then** P'
                 assert α
       else P''
                 assert β **fi**

L: **assert** $\alpha \vee \beta$

$<28>$ *backward branch rule*

if t **then** P'
                 L':
       else P''
                 L'':   **fi**
assert β

L',L'': **assert** β

4.2.4. Loop Control Rules

$<29>$ *forward loop-body rule*

assert α
loop L:
  P
  assert β
  repeat

L: **assert** $\alpha \vee \beta$

$<30>$ *backward loop-body rule*

L':
loop assert β
  P
  L'':
  repeat

L',L'': **assert** β

$<31>$ *forward loop-exit rule*

loop P'
  assert α
  until t
  L':
  P''
  repeat
L'':

L': **assert** $\alpha, \sim t$ L'': **assert** α, t

```
┌─────────────────────────────────────┐
│ <32>  backward loop-exit rule       │
├─────────────────────────────────────┤
│ loop  P'                            │
│       L :                           │
│       until t                       │
│       assert α                      │
│       P''                           │
│       repeat                        │
│ assert β                            │
├─────────────────────────────────────┤
│ L :  assert  ∼t ⊃ α, t ⊃ β          │
└─────────────────────────────────────┘
```

4.2.5. Value Rules

- $\overline{x}_L$ denotes the value of the variables $\overline{x}$ when control was last at label L. An invariant containing $\overline{x}_L$ may not be pushed over label L.

```
┌──────────────────────────────┐
│ <33>  label axiom            │
├──────────────────────────────┤
│ L :  assert  x̄ = x̄_L         │
└──────────────────────────────┘
```

```
┌──────────────────────────────────────┐
│ <34>  protected-invariant rule       │
├──────────────────────────────────────┤
│ assert  α(x̄)                         │
│ loop  L :                            │
│       P                              │
│       assert  α(x̄) ∨ x̄ = x̄_L         │
│       repeat                         │
├──────────────────────────────────────┤
│ L :  assert  α(x̄)                    │
└──────────────────────────────────────┘
```

where $\overline{x}$ are the only variables in α.

$<35a>$ *forall rule*

---
assert $\overline{x}=\overline{a}$
loop L: **assert** $\alpha(\overline{x})$
     P
     assert $\overline{x}=f(\overline{x}_L)$
     repeat

L: **assert** $(\forall\,\overline{\xi}\in\{\overline{a},f(\overline{a}),\cdots,\overline{x}\})\ \alpha(\overline{\xi})$

---

where $\overline{x}$ are the only variables in α. This is a special case of $<37a>$ below which is guaranteed to yield an invariant and not just a candidate.

$<35b>$ *integer forall rule*

---
assert $x=a$, $x\in\mathbf{Z}$
loop L: **assert** $\alpha(x)$
     P
     assert $x=x_L+1$
     repeat

L: **assert** $(\forall\,\xi\in[a:x])\ \alpha(\xi)$

---

where $\overline{x}$ are the only variables in α.

4.3. Heuristic Rules

The heuristic rules all yield local candidate assertions. The "dangerous" ones should be applied with caution, when there is additional reason to believe that they may work.

4.3.1. Control Heuristics

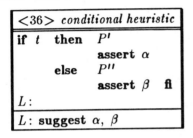

$$\boxed{\begin{array}{l} \text{<36>} \; \textit{conditional heuristic} \\ \hline \textbf{if } t \quad \textbf{then} \quad P' \\ \qquad\qquad\qquad \textbf{assert } \alpha \\ \qquad\quad \textbf{else} \quad P'' \\ \qquad\qquad\qquad \textbf{assert } \beta \quad \textbf{fi} \\ L: \\ \hline L: \textbf{suggest } \alpha, \; \beta \end{array}}$$

$$\boxed{\begin{array}{l} \text{<37a>} \; \textit{generalization heuristic} \\ \hline \textbf{assert } \overline{x} = \overline{a} \\ \textbf{loop } L: \textbf{suggest } \alpha(\overline{x}, \overline{y}) \\ \qquad P \\ \qquad \textbf{assert } \overline{x} = f(\overline{x}_L) \\ \qquad \textbf{repeat} \\ \hline L: \textbf{suggest } (\forall\, \overline{\xi} \in \{\overline{a}, f(\overline{a}), f(f(\overline{a})), \cdots, x\}) \; \alpha(\overline{\xi}, \overline{y}) \end{array}}$$

$$\boxed{\begin{array}{l} \text{<37b>} \; \textit{integer generalization heuristic} \\ \hline \textbf{assert } x = a, \; x \in \mathbf{Z} \\ \textbf{loop } L: \textbf{suggest } \alpha(x, y) \\ \qquad P \\ \qquad \textbf{assert } x = x_L + 1 \\ \qquad \textbf{repeat} \\ \hline L: \textbf{suggest } (\forall\, \xi \in [a{:}x]) \; \alpha(\xi, y) \end{array}}$$

```
┌──────────────────────────────────┐
│ <38>  top-down heuristic         │
├──────────────────────────────────┤
│ suggest α                        │
│ loop L :                         │
│       until t                    │
│       P                          │
│       repeat                     │
│ suggest γ                        │
│                                  │
│ fact γ when α                    │
├──────────────────────────────────┤
│ L : suggest γ                    │
└──────────────────────────────────┘
```

4.3.2. Dangerous Heuristics

```
┌──────────────────────────────────┐
│ <39>  disjunction heuristic      │
├──────────────────────────────────┤
│ L : suggest α ∨ β                │
├──────────────────────────────────┤
│ L : suggest α, β                 │
└──────────────────────────────────┘
```

This should be used in conjunction with $<27>$.

```
┌──────────────────────────────────┐
│ <40>  strengthening heuristic    │
├──────────────────────────────────┤
│ L :   suggest α(x̄)              │
│       suggest γ(x̄)              │
├──────────────────────────────────┤
│ L : suggest (∀ξ̄) (α(ξ̄) ⊃ γ(ξ̄)) │
└──────────────────────────────────┘
```

This should be used in conjunction with $<38>$.

```
┌──────────────────────────────────┐
│ <41>  transitivity heuristic     │
├──────────────────────────────────┤
│ L :   suggest α(x̄,ȳ)            │
│       suggest α(x̄,z̄)            │
│                                  │
│ fact α(u,w) when α(u,v) ∧ α(v,w) │
├──────────────────────────────────┤
│ L : suggest α(x̄,w̄) ∨ ȳ=z̄       │
└──────────────────────────────────┘
```

i.e. α is a transitive relation. This should be used in conjunction with $<38>$.

Appendix 5

Implementation

Intelligent (3): able to perform some of the functions of a computer.

—Webster's new collegiate dictionary

If there were a machine which ... imitated our actions as far as is morally possible, there would always be ... absolutely certain methods of recognizing that it was still not truly a man.

—René Descartes (Discourse on the method of rightly conducting the reason and seeking truth in the field of science)

5.1. Introduction

The ideas presented in this monograph have been implemented in three (loosely connected) programs:

- a *modification system* to apply analogical reasoning to program transformations,

- a *synthesis system* to construct iterative programs from input-output specifications, and

- an *annotation system* to generate invariant assertions and candidates from program text.

The three systems had parts in common, but were not otherwise joined into a unified system. They were used as a "test bed" for developing our methods.

The programs were written in QLISP ([Wilber76]), which resides in an INTERLISP environment ([Teitelman74]). QLISP provides pattern-directed procedure invocation, backtracking facilities, and convenient data types. A simplifier provides the necessary arithmetic and logical reasoning ability; it is in the form of QLISP procedures and was modeled after [WaldingerLevitt74].[1] For example, the simplification rule

$$u_1 \cdot u_2 \cdots u_n \cdot 1 \;\rightarrow\; u_1 \cdot u_2 \cdots u_n$$

is implemented by the procedure

(QLAMBDA (SIMP-TIMES-1 (TIMES ←←U 1) ('(TIMES $$U)))) ,

with QLISP taking the commutativity and associativity of "TIMES" into account automatically. There are also simplification rules for set expressions, e.g.

$$\mathbf{N + N} \;\rightarrow\; \mathbf{N}.$$

The only theorem proving is via simplification; when in doubt, the user is queried.

---

[1] Some other simplifiers incorporated in verification systems are described, for example, in [Deutsch73] and [NelsonOppen80].

The following subsections include traces of examples run on the systems. The traces have been edited to improve the format and annotated to enhance the presentation.

5.2. Modification

The implemented modifier includes the following phases:

- verification condition generation from a given annotated program;

- a rudimentary attempt to bring the specifications of the given and desired programs closer syntactically.

- matching the specifications;

- checking the validity of the transformations by applying the transformations of each possible analogy to the verification conditions;

- applying the valid transformations to the programs;

- eliminating nonprimitives and simplifying the resultant program.

Some of the necessary theorem proving is done by the system; the rest is left to the user. Debugging and instantiation are special cases of modification; abstraction was not implemented. The rules for applying transformations were given in Appendix 1.

Our system successfully modified the examples in Chapter 3 and several others in the numeric and array manipulation domains. The following is an annotated QLISP trace of part of the debugging of the real division program, as executed by our modification system.

The procedure MODIFY modifies a program to achieve a new goal. Here it is used to debug a real division program:

    MODIFY:

This is the given annotated program:

(GIVEN PROGRAM)

```
[ (ASSERT (AND (LTQ 0 C)
               (LT C (TIMES 2 D))
               (LT 0 E)))

  (SETQ Z 0)
  (SETQ Y 1)
  (LOOP [ASSERT (AND (LTQ Z (DIVIDE C D))
                     (LT (DIVIDE C D) (PLUS Z (TIMES 2 Y]
        (UNTIL (LTQ Y E))
        (IF (LTQ (TIMES D (PLUS Z Y)) C)
            THEN (SETQ Z (PLUS Z Y))
            FI)
        (SETQ Y (DIV2 Y))
        REPEAT)
  (ASSERT (AND (LTQ Z (DIVIDE C D))
               (LT (DIVIDE C (TIMES D 2))
                   (PLUS (DIVIDE Z 2) E]
```

prefaced by an input assertion, (containing the conditions under which the invariants hold), and followed by output invariants. (A square right bracket is used to close all unclosed open parentheses, as well as the corresponding left bracket.) We desire that the program achieve the specification:

(DESIRED SPECIFICATIONS)

```
[ (ASSERT (AND (LTQ 0 C)
               (LT C D)
               (LT 0 E)))
  (ACHIEVE AND (LTQ Z (DIVIDE C D))
               (LT (DIVIDE C D) (PLUS Z E]
```

The primitive operations allowed in the final program are also specified:

PRIMITIVES : (EQ LTQ GTQ LT GT ADD1 CONTENTS SUB1 ASSIGN
DIV2 MINUS PLUS SUBTRACT TIMES SQUARE)

Note that the general division operator is omitted, though division by
two (DIV2) is permitted.

The system first scanned the program to distinguishes between variables and constants appearing in it:

VARIABLES : (Z Y)
CONSTANTS : (AND LTQ 0 C LT TIMES 2 D E 1 DIVIDE PLUS DIV2)

The first step is to apply the function MATCH to compare the given output invariant with the desired output specification:

COMPARE :
A= [AND (LTQ Z (DIVIDE C D))
         (LT (DIVIDE C (TIMES D 2))) (PLUS E (DIVIDE Z 2]
B= (AND (LTQ Z (DIVIDE C D))
         (LT (DIVIDE C D)   (PLUS E Z)))

(A and B are the formal parameters of the procedure COMPARE.) Matching the second conjuncts:

     MATCH :
     G1= [AND (LT (DIVIDE C (TIMES 2 D))) (PLUS E
                                            (DIVIDE Z 2]
     G2= (AND (LT (DIVIDE C D)   (PLUS E Z)))

the system notices that if the expression (TIMES 2 D) could be transformed
into D and (DIVIDE Z 2) into Z, then the whole conjunct would transform
as desired. So the function INVERT is called; function inversion shows
that there are two ways to transform (TIMES 2 D) into D:

INVERT:
A= (TRANSFORM (TIMES 2 D) D)
(INVERT) = [(TRANSFORM 2 (DIVIDE D (TIMES D)))
              (TRANSFORM D (DIVIDE D (TIMES 2]

either (TRANSFORM 2 1) or (TRANSFORM D (DIVIDE D 2)).

Similarly, to transform (DIVIDE Z 2) into Z:

INVERT:
A= (TRANSFORM (DIVIDE Z 2) Z)
(INVERT) = ((TRANSFORM Z (TIMES 2 Z))
             (TRANSFORM 2 (DIVIDE Z Z)))

Another possibility is suggested by the commutativity of PLUS (handled automatically by QLISP):

INVERT:
A= (TRANSFORM (DIVIDE Z 2) E)
(INVERT) = ((TRANSFORM Z (TIMES 2 E))
             (TRANSFORM 2 (DIVIDE Z E)))

but is rejected, since program variables cannot be transformed into expressions containing only input variables, nor can constants be transformed into variable expressions.

At this point, a question is posed to the human theorem prover: true or false?

(T OR NIL?)
[IMPLIES [LT C (TIMES D 2 (PLUS E (DIV2 Z]
          (LT C (TIMES D (PLUS E Z]

If the answer is yes, then no transformations are necessary, since the output invariant implies the output specification. His answer, however, is negative.

: n i l

The system has succeeded in finding one satisfactory match for the second conjunct:

```
(MATCH)= [((TRANSFORM LT LT)
          (TRANSFORM DIVIDE DIVIDE)
          (TRANSFORM C C)
          (TRANSFORM D (DIVIDE D (TIMES 2))))
          (TRANSFORM PLUS PLUS)
          (TRANSFORM E E)
          (TRANSFORM Z (TIMES 2 Z]
```

Applying them to the first conjunct:

```
TRANSFORM-EXPRESSION
R= (LTQ Z (DIVIDE C D))
(TRANSFORM-EXPRESSION) = [LTQ (TIMES 2 Z)
                              (DIVIDE C
                              (DIVIDE D (TIMES 2]
```

and simplifying, proves the desired conjunct; no additional transformations are necessary.

Before proceeding, the system looks for additional possible transformations. The systems finds equivalent formulations of the specifications:

```
[AND (LTQ (TIMES Z D) C)
     (LT C (PLUS (TIMES Z D) (TIMES 2 E D]
```

and

```
[AND (LTQ (TIMES Z D) C)
     (LT C (PLUS (TIMES Z D) (TIMES E D]
```

Comparing them yields the transformation

$$[((\text{TRANSFORM E} (\text{DIVIDE E} 2]$$

Thus, two sets of transformations have been obtained:

$$(\text{COMPARE}) = [((\text{TRANSFORM D} (\text{DIVIDE D} (\text{TIMES} 2))))$$
$$(\text{TRANSFORM Z} (\text{TIMES} 2 Z)))$$
$$((\text{TRANSFORM E} (\text{DIVIDE E} 2]$$

The system now calls the function TRANSFORM-PROGRAM for each of the two eligible transformations in turn:

TRANSFORM-PROGRAM:

To transform the input variable D:

TRANSFORM-INPUT-VARIABLE:
$S= (\text{TRANSFORM D} (\text{DIVIDE D} (\text{TIMES} 2)))$

Transforming the program variable Z

TRANSFORM-PROGRAM-VARIABLE:
$S= (\text{TRANSFORM Z} (\text{TIMES} 2 Z))$

may entail eliminating expressions from the left-hand side of assignments; the function INVERT is used to compute the new right-hand sides:

INVERT:
$A= (\text{SETQ} (\text{TIMES} 2 Z) 0)$
$(\text{INVERT}) = [(\text{SETQ Z} (\text{DIVIDE} 0 (\text{TIMES} 2]$

and

```
INVERT:
A= (SETQ (TIMES 2 Z) (PLUS Y (TIMES Z 2)))
(INVERT) = [(SETQ Z
                (DIVIDE (PLUS Y (TIMES 2 Z)) (TIMES 2]
```

The transformed program is:

(TRANSFORMED PROGRAM)

```
[[ASSERT (AND (LTQ 0 C)
              (LT 0 E)
              (LT C (TIMES (DIVIDE D (TIMES 2)) 2]
 (SETQ Z (DIVIDE 0 (TIMES 2)))
 (SETQ Y 1)
 (LOOP [ASSERT (AND [LTQ (TIMES 2 Z)
                         (DIVIDE C (DIVIDE D (TIMES 2]
                    (LT (DIVIDE C (DIVIDE D (TIMES 2)))
                        (PLUS (TIMES 2 Z) (TIMES Y 2]
       (UNTIL (LTQ Y E))
       (IF (LTQ (TIMES (DIVIDE D (TIMES 2))
                (PLUS Y (TIMES Z 2))) C)
           THEN (SETQ Z
                    (DIVIDE (PLUS Y (TIMES Z 2))
                            (TIMES 2)))
           FI)
       (SETQ Y (DIV2 Y))
       REPEAT)
 (ASSERT (AND [LTQ (TIMES Z 2) (DIVIDE D (TIMES 2]
              (LT [DIVIDE C (TIMES 2 (DIVIDE D (TIMES 2]
                  (PLUS E (DIVIDE (TIMES Z 2) 2]
```

Nonexecutable statements (involving DIVIDE) are now replaced by executable ones (DIV2) as part of a simplification step.

(SIMPLIFIED PROGRAM)

```
[(ASSERT (AND (LTQ 0 C)
              (LT 0 E)
              (LT C D)))
 (SETQ Z 0)
 (SETQ Y 1)
 (LOOP (ASSERT (AND [LT C (TIMES (DIV2 D)
                                 (PLUS (TIMES 2 Z)
                                       (TIMES 2 Y]
                    (LTQ (TIMES 2 Z (DIV2 D)) C)))
       (UNTIL (LTQ Y E))
       (IF (LTQ (TIMES (DIV2 D) (PLUS Y (TIMES 2 Z))) C)
           THEN [SETQ Z (DIV2 (PLUS Y (TIMES 2 Z]
           FI)
       (SETQ Y (DIV2 Y))
       REPEAT)
 (ASSERT (AND (LTQ (TIMES Z 2 (DIV2 D)) C)
              (LT C (TIMES 2 (DIV2 D) (PLUS E Z]
```

This program is indeed executable.

Lastly, it must be proved that the transformed input assertion is implied by the given input specification

```
TRANSFORM-EXPRESSION:
R= (AND (LTQ 0 C)
        (LT 0 E)
        (LT C (TIMES 2 D)))
(TRANSFORM-EXPRESSION)= [AND (LTQ 0 C)
                             (LT 0 E)
                             (LT C (TIMES 2
                                    (DIVIDE D (TIMES 2]
```

and it is, since (TIMES 2 (DIVIDE D (TIMES 2))) is equal to D, leaving a vacuous initialization goal:

```
ACHIEVE:
G1= T
(ACHIEVE)= NIL
```

The

(DESIRED PROGRAM)

is

```
[((ASSERT (AND (LTQ 0 C)
               (LT 0 E)
               (LT C D)))
  (SETQ Z 0)
  (SETQ Y 1)
  (LOOP (ASSERT (AND [LT C (TIMES (DIV2 D)
                                  (PLUS (TIMES 2 Z)
                                        (TIMES 2 Y]
                     (LTQ (TIMES 2 Z (DIV2 D)) C)))
        (UNTIL (LTQ Y E))
        (IF (LTQ (TIMES (DIV2 D) (PLUS Y (TIMES 2 Z)))
                 C)
            THEN [SETQ Z (DIV2 (PLUS Y (TIMES 2 Z]
            FI)
        (SETQ Y (DIV2 Y))
        REPEAT)
  (ASSERT (AND (LTQ (TIMES Z 2 (DIV2 D)) C)
               (LT C (TIMES 2 (DIV2 D) (PLUS E Z]
```

Applying the second set of transformations to the program

```
TRANSFORM-PROGRAM:
S= ((TRANSFORM E (DIVIDE E 2)))
```

yields

```
((ASSERT [AND (LTQ 0 C)
               (LT C (TIMES 2 D))
               (LT 0 (DIVIDE E 2]
  (SETQ Z 0)
  (SETQ Y 1)
  (LOOP [ASSERT (AND (LTQ (TIMES D Z) C)
                     (LT C (TIMES D (PLUS Z
                                          (TIMES 2 Y]
        (UNTIL (LTQ Y (DIV2 E)))
        (IF (LTQ (TIMES D (PLUS Z Y)) C)
            THEN (SETQ Z (PLUS Z Y))
            FI)
        (SETQ Y (DIV2 Y))
        REPEAT)
  [ASSERT (AND (LTQ Z (DIVIDE C D))
               (LT (DIVIDE C (TIMES 2 D))
                   (PLUS (DIVIDE Z 2) (DIVIDE E 2]))
```

Again it must be shown that the transformed input assertion is implied by the input specification:

```
(IMPLIES (AND (LTQ 0 C) (LT C D) (LT 0 E))
         [AND (LTQ 0 C) (LT C (TIMES 2 D))
              (LT 0 (DIVIDE E 2]
```

which is indeed true, since C less than 2D is implied by C less than D and E/2 greater than 0 is equivalent to E greater than 0.

5.3. Synthesis

Assignment strategies for simple variables ($x := u$) and array variables ($A[i] := u$) were implemented, and include function inversion. The *conditional strategy* was implemented. The *splitting strategies* usually lead to loops; of the loop strategies, only the forward-loop strategy was completely implemented. The *termination strategy* was implemented for increasing or decreasing integers. (These rules appear in Appendix 3.)

There are also numerous rules for universal quantifiers, inequalities, and arrays.

We begin with a goal to find the position of the smallest element in an array:

```
(ASSERT (AND (LTQ 1 M) (MEMBER M N)))
(ACHIEVE (FORALL X IN (ARRAY A 1 M)
                (LTQ (CONTENTS A Z) X))))
```

i.e. for all elements X in the array A[1:M], X must be not less than A[Z]. The input assertion states that M is a natural number (N denotes the set of natural numbers) and is not less than 1.

The procedure ACHIEVE is called. ACHIEVE then cycles through a collection of transformations and strategies until it finds an applicable one. If the path pursued ends in failure, the program backtracks to the point of the last choice.

The first try is always to see if the goal is already known to be true; but this fails. Then the program applies rules for universal quantifiers, first trying to make the range [1:M] empty:

```
(ACHIEVE (LTQ M (SUB1 1))))
```

where (SUB1 1) is 1−1=0. This too fails, as it contradicts the input assertion.

The system then tries to reduce the range to one element:

```
(ACHIEVE (AND (LTQ (CONTENTS A Z) (CONTENTS A M))
              (EQ 1 M))))
```

But, again, failure.

So a new variable Y1 is generated, to be equal to M. The protection strategy suggests the consecutive goals:

```
(ACHIEVE (FORALL X IN (ARRAY A 1 Y1)
                      (LTQ (CONTENTS A Z) X)))
(ACHIEVE (AND (PROTECT (FORALL X IN (ARRAY A 1 Y1')
                               (LTQ (CONTENTS A Z) X)))
         (EQ M Y1')))
```

where Y1 denotes the value of the variable after the first subgoal, and Y1 denotes its value after the second.

Starting with the first, the program checks if it is known to be true; it is not. Then the quantifier rules are applied in turn. To make the range empty, it tries

```
(ACHIEVE (LTQ Y1 (SUB1 1)))
```

For this it suffices to

```
(ACHIEVE (EQ (SUB1 1) (Y1))
```

which may be achieved by an assignment statement:

```
((SETQ Y1 (SUB1 1)))
```

since SUB1 is a primitive.

The second subgoal is now taken up.

```
(ACHIEVE (AND (EQ M Y1')))
              (PROTECT (FORALL X IN (ARRAY A 1 Y1')
                               (LTQ (CONTENTS A Z) X)))
```

The forward loop strategy generates a loop-body subgoal and the termination strategy determines that Y1 is increasing from 0 to M. The following facts are asserted for the loop-body subgoal:

1)   the loop invariant

```
(ASSERT (FORALL X IN (ARRAY A 1 Y1')
               (LTQ (CONTENTS A Z) X)))
```

where Y1' stands for the value of Y1 before executing the loop body;

2) the negation of the exit test for that value

```
(ASSERT (NOT (EQ M Y1')))
```

3) the fact that the exit test holds for the final value of Y1, denoted by Y1*,

```
(ASSERT (EQ M Y1*))
```

4) and the fact that the current value of Y1, Y1', lies between its initial value Y1 and final value Y1*

```
(ASSERT (AND (LTQ Y1 Y1')
              (LTQ Y1' Y1*)))
```

The loop-body subgoal is

```
(ACHIEVE (FORALL X IN (ARRAY A 1 Y1'')
                 (LTQ (CONTENTS A Z') X)))
```

where Y1'' and Z' are the post loop-body values of Y1 and Z, respectively. The first step now is to split the range of the loop-body subgoal into two parts, one is matched with what is already asserted in the loop invariant; the remainder is

```
[ACHIEVE  (AND  (LTQ  (ADD1  Y1')  Y1'')
                (LTQ  (SUB1  1)  Y1')
                (FORALL  X  IN  (ARRAY  A  (ADD1  Y1')
                                           Y1'')
                        (LTQ  (CONTENTS  A  Z')  X]
```

Using the protection strategy, the goal

```
(ACHIEVE  (FORALL  X  IN  (ARRAY  A  (ADD1  Y1')
                                     Y1'')
                  (LTQ  (CONTENTS  A  Z')  X)))
```

is taken up first, to be followed later by

```
[ACHIEVE  (AND  (LTQ  (ADD1  Y1')  Y1'')
                (LTQ  (SUB1  1)  Y1')
                (PROTECT
                 (FORALL  X  IN
                    (ARRAY  A  (ADD1  Y1')  Y1'')
                    (LTQ  (CONTENTS  A  Z')  X]
```

An attempt is made to achieve this quantified goal vacuously, i.e.

```
[ACHIEVE  (LTQ  Y1''  (SUB1  (ADD1  Y1']
```

To achieve this inequality, the system tries the sufficient

```
(ACHIEVE  (EQ  (SUB1  (ADD1  Y1'))  Y1''))
```

This may indeed be achieved by leaving the value of Y1 unchanged, but then the remaining conjunct

```
(ACHIEVE  (LTQ  (ADD1  Y1')  Y1'''))
```

cannot be achieved.

Backtracking to the quantified goal

$$(ACHIEVE \ (FORALL \ X \ IN \ (ARRAY \ A \ (ADD1 \ Y1\,') \\ Y1\,'\,') \\ (LTQ \ (CONTENTS \ A \ Z\,') \ X)))$$

and attempting to collapse the range into a single element, rather than the empty range, gives the conjunctive goal

$$(ACHIEVE \ (AND \ (LTQ \ (CONTENTS \ A \ Z\,') \\ (CONTENTS \ A \ Y1\,'\,')) \\ (EQ \ (ADD1 \ Y1\,') \ Y1\,'\,')))$$

The conjunct

$$(ACHIEVE \ (EQ \ (ADD1 \ Y1\,') \ Y1\,'\,'))$$

may be achieved by an assignment statement

$$((SETQ \ Y1 \ (ADD1 \ Y1)))$$

leaving

$$[ACHIEVE \ (AND \ (LTQ \ (CONTENTS \ A \ Z\,') \\ (CONTENTS \ A \ Y1\,'\,'\,')) \\ (PROTECT \ (EQ \ (ADD1 \ Y1\,') \\ Y1\,'\,'\,']$$

where Y1''' denotes the value of Y1 after this goal.

Protecting the relation (EQ (ADD1 Y1') Y1'') by not varying Y''', i.e. Y1'''=Y'', gives

$$(ACHIEVE \ (LTQ \ (CONTENTS \ A \ Z\,') \\ (CONTENTS \ A \ Y1\,'\,')))$$

which strengthens to

$$(\text{ACHIEVE} \ (\text{EQ} \ (\text{CONTENTS} \ A \ Z') \\ (\text{CONTENTS} \ A \ Y1'')))$$

This in turn suggests

$$(\text{ACHIEVE} \ (\text{EQ} \ Y1'' \ Z'))$$

where Z' is a variable and may be assigned to

$$((\text{SETQ} \ Z \ Y1))$$

That leaves

$$[\text{ACHIEVE} \ (\text{AND} \ (\text{LTQ} \ (\text{ADD1} \ Y1') \ Y1'') \\ (\text{PROTECT} \\ (\text{FORALL} \ X \ \text{IN} \\ (\text{ARRAY} \ A \ (\text{ADD1} \ Y1') \ Y1'') \\ (\text{LTQ} \ (\text{CONTENTS} \ A \ Z'') \ X]$$

The conjunct

$$(\text{ACHIEVE} \ (\text{AND} \ (\text{LTQ} \ (\text{ADD1} \ Y1') \ Y1'')))$$

is proved to be true for the new value of Y1, however, the remaining conjunct

$$[\text{ACHIEVE} \ (\text{PROTECT} \\ (\text{FORALL} \ X \ \text{IN} \ (\text{ARRAY} \ A \ 1 \ Y1') \\ (\text{LTQ} \ (\text{CONTENTS} \ A \ Z') \ X]$$

cannot be proved.

So again the system backtracks and tries forming a conditional to

$$(\text{ACHIEVE} \;\; (\text{LTQ} \;\; (\text{CONTENTS} \;\; A \;\; Z') \\ (\text{CONTENTS} \;\; A \;\; Y1'')))$$

This leads to two subgoals, corresponding to whether or not (LTQ (CON-TENTS A Z) (CONTENTS A Y1)) holds. If it does, then

$$(\text{ASSERT} \;\; (\text{LTQ} \;\; (\text{CONTENTS} \;\; A \;\; Z) \\ (\text{CONTENTS} \;\; A \;\; Y1'')))$$

and attempt to

$$[\text{ACHIEVE} \;\; (\text{AND} \;\; (\text{LTQ} \;\; (\text{ADD1} \;\; Y1') \;\; Y1'') \\ (\text{PROTECT} \\ (\text{FORALL} \;\; X \;\; \text{IN} \\ (\text{ARRAY} \;\; A \;\; (\text{ADD1} \;\; Y1') \;\; Y1'') \\ (\text{LTQ} \;\; (\text{CONTENTS} \;\; A \;\; Z) \;\; X]$$

This time, each conjunct in turn is found to be true, and the empty program is returned:

$$(\text{ASSERT-ACHIEVE}) \;\; = \;\; \text{NIL}$$

Taking up the second subgoal, corresponding to when the test is false,

$$[\text{ASSERT} \;\; (\text{NOT} \;\; (\text{LTQ} \;\; (\text{CONTENTS} \;\; A \;\; Z) \\ (\text{CONTENTS} \;\; A \;\; Y1'']$$

the system attempts to

$$(\text{ACHIEVE} \;\; (\text{LTQ} \;\; (\text{CONTENTS} \;\; A \;\; Z') \\ (\text{CONTENTS} \;\; A \;\; Y1'')))$$

Again, this suggests

$$(ACHIEVE\ (EQ\ (CONTENTS\ A\ Y1''\,)$$
$$(CONTENTS\ A\ Z'\,)))$$

which in turn suggests

$$(ACHIEVE\ (EQ\ Y1''\ Z'\,))$$

which is achieved by an assignment statement

$$((SETQ\ Z\ Y1))$$

The remaining conjuncts are proved true, one by one.

$$(ACHIEVE\ (AND\ (LTQ\ (ADD1\ Y1'\,)\ Y1''\,)))$$

is found to hold, leaving

$$(ACHIEVE\ (FORALL\ X\ IN\ (ARRAY\ A\ 1\ Y1'\,)$$
$$(LTQ\ (CONTENTS\ A\ Z'\,)$$
$$X)))$$

which is first transformed into

$$(ACHIEVE\ (LTQ\ (CONTENTS\ A\ Z'\,)$$
$$(ARRAY\ A\ 1\ Y1'\,)))$$

so that the QLISP facility for handling transitive relations may be of help. Based on the asserted relations, this leads to the subgoal

$$(ACHIEVE\ (LTQ\ (CONTENTS\ A\ Z'\,)$$
$$(CONTENTS\ A\ Z)))$$

This leads to

$$(\text{ACHIEVE} \ (\text{LTQ} \ (\text{CONTENTS} \ A \ Z\,') \\ (\text{CONTENTS} \ A \ Y1\,'\,')))$$

which is found to be true:

$$(\text{PROVE}) \ = \ T$$

Now the protected conjuncts

$$(\text{ACHIEVE} \ (\text{AND} \ (\text{LTQ} \ (\text{ADD1} \ Y1\,') \ Y1\,'\,') \\ (\text{LTQ} \ Y1\,'\,' \ Y1\,'*)))$$

are also found to hold, and the empty program is returned:

$$(\text{PROTECT}) \ = \ \text{NIL}$$

Recapitulating, for the false path of the conditional, the result is

$$(\text{ASSERT-ACHIEVE}) \ = \ ((\text{SETQ} \ Z \ Y1))$$

the conditional is

$$(\text{ACHIEVE-IF}) \ = \ ((\text{IF} \ (\text{LTQ} \ (\text{CONTENTS} \ A \ Z) \\ (\text{CONTENTS} \ A \ Y1)) \\ \text{THEN} \\ \text{ELSE} \ (\text{SETQ} \ Z \ Y1) \\ \text{FI}))$$

and together with the assignment to Y1, the loop-body is

```
(ACHIEVE-SETQ) = ((SETQ Y1 (ADD1 Y1))
                  (IF (LTQ (CONTENTS A Z)
                           (CONTENTS A Y1))
                      THEN
                      ELSE (SETQ Z Y1)
                  FI))
```

The whole loop is

```
(ACHIEVE-LOOP) = ((LOOP [ASSERT
                        (AND (LTQ Y1 Y1')
                             (LTQ Y1' Y1*)
                             (FORALL X IN
                                 (ARRAY A 1 Y1)
                                 (LTQ (CONTENTS A Z)
                                      X]
                        (UNTIL (EQ M Y1))
                        (SETQ Y1 (ADD1 Y1))
                        (IF (LTQ (CONTENTS A Z)
                                 (CONTENTS A Y1))
                            THEN
                            ELSE (SETQ Z Y1)
                        FI)
                  REPEAT))
```

and with the initialization, the program is

```
(ACHIEVE-SETQ) =

    ((SETQ Y1 (SUB1 1))
     (LOOP [ASSERT
             (AND (LTQ Y1 Y1')
                  (LTQ Y1' Y1*)
                  (FORALL X IN (ARRAY A 1 Y1)
                               (LTQ (CONTENTS A Z)
                                    X]
           (UNTIL (EQ M Y1))
           (SETQ Y1 (ADD1 Y1))
           (IF (LTQ (CONTENTS A Z)
                    (CONTENTS A Y1))
               THEN
               ELSE (SETQ Z Y1)
               FI)
           REPEAT))
```

Thus, the final program is

```
[(SETQ Y1 (SUB1 1))
 (LOOP [ASSERT (AND (LTQ Y1 Y1') (LTQ Y1' Y1*)
               (FORALL X IN (ARRAY A 1 Y1)
                            (LTQ (CONTENTS A Z) X]
       (UNTIL (EQ M Y1))
       (SETQ Y1 (ADD1 Y1))
       (IF (LTQ (CONTENTS A Z) (CONTENTS A Y1))
           THEN
           ELSE (SETQ Z Y1)
           FI)
       REPEAT]
```

The initialization assignment

```
(SETQ Y1 (SUB1 1))
```

could be simplified to

(SETQ Y1 0)

Note that Z is not initialized before entering the loop. This is because the specifications did not include (LTQ 1 Z N).

5.4. Annotation

Our annotation system allows the user access to both high level routines and individual rules. Thus the user has the option of interacting with the system and guiding the order of rule application. The rules may be found in Appendix 4.

Bounds on variables are derived from range invariants (rules $<1>$-$<7>$) using methods adapted from "interval arithmetic" (see, for example [Gibb61] and [Yohe79]), a form of "weak interpretation" (cf. [Sintzoff72]). For example, if we know that the variable x lies within the range $[a:b]$ and y lies within $[c:d]$, then it follows that $x+y$ has the range $[a+c:b+d]$. Rules $<8>$-$<10>$ are difference equation solutions using counters and have been implemented by [Tamir80]. The matching routine used for discovering analogies between specifications, may also be used to recognize instances of the global relation rules $<11>$-$<17>$.

The control rules have been implemented as an "operator grammar"([Gerhart75c]). The *forward axioms* $<18>$-$<20>$ derive assertions from the text. The *forward assignment rule* $<21>$ uses function inversion (cf. [Elspas74]); array assignments are handled by rules $<23>$ and $<24>$. When a function is not invertible, as much information as possible is nevertheless derived. The remainder of the *forward rules* ($<27>$-$<31>$) push assertions forward over program paths. *Backward control rules* ($<22>$-$<28>$) push assertions backwards over program paths and may also be used to generate the program's verification conditions.

The *conditional* and *generalization heuristics* ($<36>$ and $<37>$) are applied in conjunction with the *forward control rules*. The *top-down heuristic* ($<38>$) is used with the *backward control rules*, with the choice of candidate chosen by a set of domain-dependent strengthening rules.

The transitive closure of a set of conjuncts is useful for the simplification of disjunctive invariants and strengthening of candidates. (Algorithms for handling inequalities may be found, for example, in [Eve75] and [NelsonOppen80].) Apart from limited capabilities of the simplifier, we have not implemented a theorem prover. Thus, candidates are only suggested by our system, but are not verified.

The following is a trace of the automatic annotation of a real-division algorithm. This is the corrected version of the example in Section 6.3.1, extended with a loop counter. Annotating more complex programs requires some degree of user interaction.

The given program is

```
((ASSERT (AND (LTQ 0 C)
              (LT C D)
              (LT 0 E)))
 (SETQ Q 0)
 (SETQ QQ 0)
 (SETQ S 1)
 (SETQ SS D)
 (SETQ N1 0)
 (LOOP (SUGGEST T)
       (ASSERT T)
       (UNTIL (LTQ S E))
       (SETQ S (DIVIDE S 2))
       (SETQ SS (DIVIDE SS 2))
       (IF (LTQ (PLUS QQ SS) C)
            THEN (SETQ Q (PLUS Q S))
                 (SETQ QQ (PLUS QQ SS)))
            FI)
       (SETQ N1 (PLUS N1 1))
       REPEAT)
 [SUGGEST (AND (LTQ Q (DIVIDE C D))
               (LT (DIVIDE C D) (PLUS Q E]
 (ASSERT T))
```

It computes the quotient of C and D. The variable N1 is a loop counter.

First the procedure GLOBAL is called to generate global invariants from the program's assignment statements. It, in turn, calls the procedure SINGLE for each program variable. SINGLE tries out the various assignment rules.

SINGLE:
Y= N1
(SINGLE) = [(MEMBER N1 (PLUS (UNION 0) (SIGMA (PLUS 1]

where MEMBER is "set membership" and (SIGMA x) is the set of finite sums of terms of the form x. Similarly

    [(MEMBER S (TIMES (UNION 1)
                      (PI (TIMES (RECIP 2]

    [(MEMBER SS (TIMES (PI (TIMES (RECIP 2))))
                       (UNION D]

    [(MEMBER Q (PLUS (UNION 0)
                     (SIGMA
                      (PLUS
                       (TIMES (RECIP 2)
                              (TIMES (UNION 1)
                               (PI (TIMES
                                    (RECIP 2]

    [(MEMBER QQ (PLUS (UNION 0)
                      (SIGMA
                       (PLUS
                        (TIMES (RECIP 2)
                               (TIMES
                                (PI
                                 (TIMES
                                  (RECIP 2)))
                                (UNION D]

where RECIP denotes "reciprocal" and (PI x) is the set of finite products of factors of the form x.

Then GLOBAL calls DOUBLE for pairs of variables.

```
DOUBLE:
X= N1
Y= S
(DOUBLE) = ((EQ (TIMES (EXP (PLUS 0
                                (TIMES S
                                  (SUBTRACT (RECIP 2)
                                            1))))
                      1)
                (EXP (RECIP 2) 0))
           (TIMES (EXP (PLUS 0
                             (TIMES 1
                               (SUBTRACT (RECIP 2)
                                         1)))
                       1)
                  (EXP (RECIP 2) N1)))
```

Similarly

```
            ((EQ (TIMES (EXP (RECIP 2) 0)
                        (EXP (PLUS 0
                                   (TIMES SS
                                     (SUBTRACT (RECIP 2)
                                               1)))
                             1))
                 (TIMES (EXP (RECIP 2) N1)
                        (EXP (PLUS 0
                                   (TIMES D
                                     (SUBTRACT (RECIP 2)
                                               1)))
                             1)))
```

```
[(EQ [TIMES (EXP (PLUS 0
                       (TIMES S
                          (SUBTRACT (RECIP 2)
                                    1)))
                  (LOG (RECIP 2)))
            (EXP (PLUS 0
                       (TIMES D
                          (SUBTRACT (RECIP 2)
                                    1)))
                 (LOG (RECIP 2)]
     (TIMES (EXP (PLUS 0
                       (TIMES SS
                          (SUBTRACT (RECIP 2)
                                    1)))
                 (LOG (RECIP 2)))
            (EXP (PLUS 0
                       (TIMES 1
                          (SUBTRACT (RECIP 2)
                                    1)))
                 (LOG (RECIP 2)]
```

```
[(EQ (TIMES .5 (SUBTRACT QQ 0))
     (TIMES (TIMES .5 D) (SUBTRACT Q 0]
```

The global invariants are then simplified:

```
(GLOBAL) - 1 =
    (AND (MEMBER N1 N)
         (MEMBER S (EXP .5 N))
         (MEMBER SS (TIMES D (EXP .5 N)))
         [MEMBER Q (SIGMA (EXP .5 (PLUS 1 N]
         [MEMBER QQ (TIMES D (SIGMA (EXP .5 (PLUS 1 N]
         (EQ S (EXP .5 N1))
         (EQ SS (TIMES D (EXP .5 N1)))
         (EQ SS (TIMES D S))
         (EQ QQ (TIMES D Q)))
```

From these invariants, the value of variables may be derived, e.g.

VALUE(Q)

returns

(DIVIDE QQ D)

The range of some of the variables may also be computed

RANGE(S)

yields

(LTQ (RECIP W) S 1.0)

where W stands for "infinity" and (RECIP W) is infitesmal; thus, this invariant means (LT 0 S) and (LTQ S 1).

Finally, the system applies the control rules and conditional heuristic. All together, of invariants (1-13) of Section 6.3.1, the system automatically generated all except (3,4,13), which follow from the generated invariants, but require more sophisticated knowledge to obtain explicitly. Note that several non-invariant candidates were also suggested by the system.

The final output is

```
((ASSERT (AND (LTQ 0 C) (LT C D) (LT 0 E)))
 (SUGGEST T)
 (SETQ Q 0) (SETQ QQ 0) (SETQ S 1) (SETQ SS D) (SETQ N1 0)
 (LOOP [ASSERT (OR (AND (EQ 0 N1) (EQ SS D) (EQ 1 S)
                        (EQ QQ N1) (EQ Q QQ) (LTQ Q C)
                        (LT C SS) (LT Q E))
                   (AND (GT (TIMES 2 S) E)
                        (LTQ QQ C))
                   (AND (GT (TIMES 2 S) E)
                        (GT (PLUS QQ SS) C]
         (SUGGEST (AND (EQ 0 N1) (EQ SS D) (EQ 1 S)
                       (EQ QQ N1) (EQ Q QQ)
                       (LTQ Q C)
                       (LT C SS)
                       (LT Q E)
                       (LTQ QQ C)
                       (GT (TIMES 2 S) E)
                       (GT (PLUS QQ SS) C)))
         (UNTIL (LTQ S E))
         (SETQ S (DIVIDE S 2))
         (SETQ SS (DIVIDE SS 2))
         (IF (LTQ (PLUS QQ SS) C)
             THEN (SETQ Q (PLUS Q S))
                  (SETQ QQ (PLUS QQ SS))
             FI)
         (SETQ N1 (PLUS N1 1))
         REPEAT)
 (SUGGEST [AND (LTQ Q (DIVIDE C D))
               (LT (DIVIDE C D) (PLUS Q E])
 [ASSERT (AND (LTQ S E)
              (OR (AND (EQ 0 N1) (EQ SS D) (EQ 1 S)
                       (EQ QQ N1) (EQ Q QQ)
                       (LTQ Q C) (LT C SS) (LT Q E))
                  (AND (GT (TIMES 2 S) E) (LTQ QQ C))
                  (AND (GT (TIMES 2 S) E)
                       (GT (PLUS QQ SS) C]
 (SUGGEST T))
```

Only some of the generated loop candidates are in fact invariant. The control rules may be used to push them to the end of the program.

References

He who relates something in the name of its author, brings redemption to the world.

— Talmud

[Ada81]
United States Department of Defense,
The Programming Language Ada: Reference Manual
Springer-Verlag, Berlin, 1981

[AdamLaurent80]
Adam, A. and Laurent, J. P.
LAURA, a system to debug student programs
Artificial Intelligence **15**, pp. 75-122, 1980

[Allen69]
Allen, F. E.
Program Optimization
In C. J. Shaw, editor, *Annual Review in Automatic Programming*,
vol. 5, pp. 239-308
Pergamon Press, Oxford, 1969

[AllenCocke72]
Allen, F. E. and Cocke, J.
A catalogue of optimizing transformations
In R. Rustin, editor, *Design and Optimization of Compilers*, pp. 1-30
Prentice-Hall, Englewood Cliffs, NJ, 1972

[Arsac79]
Arsac, J. J.
Syntactic source to source transforms and program manipulation
Comm. ACM **22**(1), pp. 43-54, Jan. 1979

[ArsacKodratoff82]
Arsac, J. J. and Kodratoff, Y.
Some techniques for recursion removal from recursive functions
ACM Trans. on Programming Languages and Systems **4**(2), pp. 295-322, April 1982

[BackManillaRaiha83]
Back, R. J. R., Manilla, H., and Raiha, K. J.
Derivation of efficient dag marking algorithms
Proc. 10th Symp. on Principles of Programming Languages, Austin,
TX, pp. 20-27, Jan. 1983

[Badger,*etal*.82]
Badger, G. F., Campbell, R. H., Dershowitz, N., Harandi, M. T.,
Laursen, A. L., Michalski, R. S., Michie, D., Penka, R., and Simmonds, M.
Knowledge based programming assistant
Report DCS-F-82-894, Dept. of Computer Science, Univ. of Illinois,
Urbana, IL, April 1982

[Balzer81]
   Balzer, R. M.
   Transformational implementation: An example
   IEEE Trans. Software Engineering **SE-7**(1), pp. 3-14, Jan. 1981

[BalzerGoldmanWile76]
   Balzer, R. M., Goldman, N., and Wile, D.
   On the transformational implementation approach to programming
   Proc. 2nd Intl. Conf. on Software Engineering, San Francisco, CA,
   pp. 337-344, Oct. 1976

[BalzerGoldmanWile77]
   Balzer, R. M., Goldman, N., and Wile, D.
   Informality in program specifications
   Proc. 5th Intl. Joint Conf. on Artificial Intelligence, Cambridge, MA,
   pp. 389-397, Aug. 1977

[Barstow79]
   Barstow, D. R.
   Knowledge-based Program Construction
   Elsevier North-Holland, New York, NY, 1979

[BasuMisra75]
   Basu, S. K. and Misra, J.
   Proving loop programs
   IEEE Trans. Software Engineering **SE-1**(1), pp. 76-86, Mar. 1975

[Bauer76]
   Bauer, F. L.
   Programming as an evolutionary process
   Proc. 2nd Intl. Conf. on Software Engineering, San Francisco, CA,
   pp. 223-234, Oct. 1976

[Bauer79]
   Bauer, M. A.
   Programming by examples
   Artificial Intelligence **12**(1), pp. 1-21, May 1979

[Bibel80]
Bibel, W.
Syntax-directed, semantics-supported program synthesis
Artificial Intelligence **14**(3), pp. 243-262, Oct. 1980

[Biermann76]
Biermann, A. W.
Approaches to automatic programming
In *Advances in Computers*, vol. 15, pp. 1-63
Academic Press, New York, NY, 1976

[Biermann78]
Biermann, A. W.
The inference of regular Lisp programs from examples
IEEE Trans. Systems, Man, and Cybernetics **SMC-8**, pp. 585-600,
Aug. 1978

[Biermann81]
Biermann, A. W.
Natural language programming
*Proc. NATO Adv. Studies Inst. on Computer Program Systems
Methodologies*, Bonas, France, pp. 335-368, Sept. 1981

[BiermannBaumPetry75]
Biermann, A. W., Baum, R. I., and Petry, F. E.
Speeding up the synthesis of programs from traces
IEEE Trans. Computers **C-24**, pp. 122-136, 1975

[BiermannKrishnaswamy76]
Biermann, A. W. and Krishnaswamy, R.
Constructing programs from example computations
IEEE Trans. Software Engineering **SE-2**(3), pp. 141-153, Sept. 1976

[BiggerstaffJohnson77]
Biggerstaff, T. J. and Johnson, D. L.
Design directed program synthesis
Technical Report 77-02-01, Dept. of Computer Science, Univ. of
Washington, Seattle, WA, Feb. 1977

[Bird77]
Bird, R. S.
Notes on recursion removal
Comm. ACM **20**(6), pp. 434-439, June 1977

[Blikle78]
Blikle, A.
Towards mathematical structured programming
In E. J. Neuhold, editor, *Formal Descriptions of Programming Concepts*, pp. 9.1-9.19
North-Holland, 1978

[BoyerElspasLevitt75]
Boyer, R. S., Elspas, B., and Levitt, K. N.
SELECT—A formal system for testing and debugging programs by symbolic execution
Proc. Intl. Conf. on Reliable Software, Los Angeles, CA, pp. 234-245, Apr. 1975

[BoyerMoore80]
Boyer, R. S. and Moore, J S.
Computational Theorem Proving
Academic Press, New York, NY, 1980

[Brotsky81]
Brotsky, D.
Program understanding through cliche recognition
Working paper 224, Artificial Intelligence Lab., Mass. Inst. of Technology, Cambridge, MA, Dec. 1981

[BrownTarnlund79]
Brown, F. M. and Tarnlund, S. A.
Inductive reasoning on recursive equations
Artificial Intelligence **12**(3), pp. 207-229, Nov. 1979

[Brown76]
Brown, R. H.
Reasoning by analogy
Working paper 132, Artificial Intelligence Lab., Mass. Inst. of Technology, Cambridge, MA, Oct. 1976

[Brown81]
Brown, R. H.
Automatic synthesis of numerical computer programs
Proc. 7th Intl. Joint Conf. on Artificial Intelligence, Vancouver, Canada, pp. 998-1003, Aug. 1981

[BroyKriegBruckner80]
Broy, M. and Krieg-Bruckner, B.
Derivation of invariant assertions during program development by transformation
ACM Trans. on Programming Languages and Systems 2(3), pp. 321-337, July 1980

[BroyPepper81]
Broy, M. and Pepper, P.
Program development as a formal activity
IEEE Trans. Software Engineering SE-7(1), pp. 14-22, Jan. 1981

[BuchananLuckham74]
Buchanan, J. R. and Luckham, D. C.
On automating the construction of programs
Memo AIM-236, Artificial Intelligence Lab., Stanford Univ., Stanford, CA, Mar. 1974

[BurstallDarlington77]
Burstall, R. M. and Darlington, J.
A transformation system for developing recursive programs
J. ACM 24(1), pp. 44-67, Jan. 1977

[Burstein83]
> Burstein, M. H.
> Concept formation by incremental reasoning and debugging
> *Proc. Intl. Machine Learning Workshop*, Monticello, IL, pp. 19-25,
> June 1983

[Caplain75]
> Caplain, M. C.
> Finding invariant assertions for proving programs
> *Proc. Intl. Conf. on Reliable Software*, Los Angeles, CA, pp. 165-171,
> Apr. 1975

[Carbonell83]
> Carbonell, J. G.
> Learning by analogy: Formulating and generalizing plans from past
> experience
> In T. M. Mitchell, editor, *Machine Learning, An Artificial Intelligence Approach*, pp. 137-162
> Tioga, Palo Alto, CA, 1983

[CheathamHollowayTownley79]
> Cheatham, T. E., Holloway, G. H., and Townley, J. A.
> Symbolic evaluation and the analysis of programs
> *IEEE Trans. Software Engineering* **SE-4**(4), pp. 402-417, July 1979

[CheathamTownley76]
> Cheatham, T. E. and Townley, J. A.
> Symbolic evaluation of programs: A look at loop analysis
> *Proc. ACM Symp. on Symbolic and Algebraic Computation*, Aug.
> 1976

[ChenFindler76]
> Chen, D. T. W. and Findler, N. V.
> Toward analogical reasoning in problem solving by computers
> Technical Report 115, Dept. of Computer Science, State Univ. of
> New York, Buffalo, NY, Dec. 1976

[Clark81]
 Clark, K. L.
 The synthesis and verification of logic programs
 Research Report DOC 81/36, Dept. of Computing, Imperial Col.,
 London, England, Sept. 1981

[Clarke76]
 Clarke, L. A.
 A system to generate test data and symbolically execute programs
 IEEE Trans. Software Engineering **SE-2**, pp. 215-222, Sept. 1976

[CockeKennedy77]
 Cocke, J. and Kennedy, K. W.
 An algorithm for reduction of operator strength
 Comm. ACM **20**(11), pp. 850-856, Nov. 1977

[ConwayGries75]
 Conway, R. and Gries, D.
 An introduction to programming: A structured approach
 Winthrop, Cambridge, MA, 1975
 (Second edition)

[Cooper66]
 Cooper, D. C.
 The equivalence of certain computations
 Comp. J. **9**, pp. 45-52, 1966

[CousotCousot77]
 Cousot, P. M. and Cousot, R.
 Automatic synthesis of optimal invvariant assertions: Mathematical
 foundations
 *Proc. ACM Symp. on Artificial Intelligence and Programming
 Languages*, Rochester, NY, pp. 1-12, Aug. 1977

[DahlDijkstraHoare72]
 Dahl, O. J., Dijkstra, E. W., and Hoare, C. A. R.
 Structured Programming
 Academic Press, London, 1972

[Darlington73]
> Darlington, J. L.
> Automatic program synthesis in second-order logic
> *Adv. Papers 3rd Intl. Joint Conf. on Artificial Intelligence*, Stanford, CA, pp. 479-485, Aug. 1973

[Darlington75]
> Darlington, J.
> Applications of program transformation to program synthesis
> *Colloques IRIA on Proving and Improving Programs*, Arc-et-Senans, France, pp. 133-144, July 1975

[Darlington78]
> Darlington, J.
> A synthesis of several sorting algorithms
> *Acta Informatica* **11**, pp. 1-30, 1978

[Darlington81]
> Darlington, J.
> An experimental program transformation and synthesis system
> *Artificial Intelligence* **16**(1), pp. 1-46, 1981

[DarlingtonBurstall76]
> Darlington, J. and Burstall, R. M.
> A system which automatically improves programs
> *Acta Informatica* **6**(1), pp. 41-60, Mar. 1976

[DeMilloLiptonPerlis79]
> De Millo, R. A., Lipton, R. J., and Perlis, A. J.
> Social processes and proofs of theorems and programs
> *Comm. ACM* **22**(5), pp. 271-280, May 1979

[DegliMiglioliOrnaghi74]
> Degli Antoni, G., Miglioli, P. A., and Ornaghi, M.
> Top-down approach to the synthesis of programs
> *Proc. Programming Symp.*, Paris, France, pp. 88-108, Apr. 1974

[Dershowitz81]
Dershowitz, N.
The evolution of programs: Program abstraction and instantiation
Proc. 5th Intl. Conf. on Software Engineering, San Diego, CA, pp. 79-88, Mar. 1981

[Dershowitz83]
Dershowitz, N.
Computing with rewrite rules
Technical Report ATR-83(8478)-1, Information Sciences Research Ofc., The Aerospace Corp., El Segundo, CA, Jan. 1983

[DershowitzManna75]
Dershowitz, N. and Manna, Z.
On automating structured programming
Colloques IRIA on Proving and Improving Programs, Arc-et-Senans, France, pp. 167-193, July 1975

[DershowitzManna77]
Dershowitz, N. and Manna, Z.
The evolution of programs: Automatic program modification
IEEE Trans. Software Engineering **SE-3**(6), pp. 377-385, Nov. 1977

[DershowitzManna81]
Dershowitz, N. and Manna, Z.
Inference rules for program annotation
IEEE Trans. Software Engineering **SE-7**(2), pp. 207-222, Mar. 1981

[Deussen79]
Deussen, P.
One abstract accepting algorithm for all kinds of parsers
Proc. 6th Intl. Colloq. on Automata, Languages and Programming, Graz, Austria, pp. 203-217, July 1979

[Deutsch73]
Deutsch, L. P.
An Interactive Program Verifier
Ph.D. Dissertation, Univ. of California, Berkeley, CA, May 1973
(Memo CSL-73-1, Xerox Research Center, Palo Alto, CA)

[Dijkstra68]
> Dijkstra, E. W.
> A constructive approach to the problem of program correctness
> *BIT* **8**(3), pp. 174-186, 1968

[Dijkstra72]
> Dijkstra, E. W.
> Notes on structured programming
> In Dahl, Dijkstra, and Hoare, *Structured Programming*, pp. 1-82
> Academic Press, London, 1972

[Dijkstra76]
> Dijkstra, E. W.
> *A Discipline of Programming*
> Prentice Hall, Englewood Cliffs, NJ, 1976

[Dijkstra78]
> Dijkstra, E. W.
> Why naive transformation systems are unlikely to work
> Working paper, Burroughs, Nuenen, The Netherlands, 1978

[DuncanYelowitz79]
> Duncan, A. G. and Yelowitz, L.
> Studies in abstract/concrete mappings in proving algorithm correctness
> *Proc. 6th EATCS Intl. Colloq. on Automata, languages and Programming*, Graz, Austria, pp. 218-229, July 1979

[Duran75]
> Duran, J. W.
> *A Study of Loop Invariants and Automatic Program Synthesis*
> Ph.D. Dissertation, Univ. of Texas, Austin, TX, May 1975
> (Report SESLTR-12, Software Engineering and Systems Lab.)

[Ellozy81]
> Ellozy, H. A.
> The determination of loop invariants for programs with arrays
> *IEEE Trans. Software Engineering* **SE-7**(2), pp. 197-206, Mar. 1981

[Elspas74]

Elspas, B.

The semiautomatic generation of inductive assertions for proving program correctness

Interim Report Project 2686, Stanford Research Institute, Menlo Park, CA, July 1974

[Elspas,*etal.*72]

Elspas, B., Green, M., Levitt, K. N., and Waldinger, R. J.

Research in interactive program proving techniques

Report, Stanford Research Institute, Menlo Park, CA, May 1972

[Evans68]

Evans, T. G.

A program for the solution of geometric-analogy intelligence test questions

In M. L. Minsky, editor, *Semantic Information Processing*, pp. 271-353

MIT Press, Cambridge, MA, 1968

[Eve75]

Eve, J.

On computing the transitive closure of a relation

Memo CS-75-508, Computer Science Dept., Stanford Univ., Stanford, CA, Sept. 1975

[Feather82]

Feather, M. S.

A system for assisting program transformation

ACM Trans. on Programming Languages and Systems **4**(1), pp. 1-20, Jan. 1982

[FikesHartNilsson72]

Fikes, R. E., Hart, P. E., and Nilsson, N. J.

Learning and executing generalized robot plans

Artificial Intelligence **3**(4), pp. 251-288, Winter 1972

[Floyd62]
>   Floyd, R. W.
>   Algorithm 113: Treesort
>   *Comm. ACM* **5**(8), p. 434, Aug. 1962

[Floyd67]
>   Floyd, R. W.
>   Assigning meanings to programs
>   *Proc. Symp. in Applied Mathematics* **19**, Providence, RI, pp. 19-32, 1967

[Floyd71]
>   Floyd, R. W.
>   Toward interactive design of correct programs
>   *Proc. Information Processing Cong.*, Ljubljana, Yugoslavia, pp. 7-10, Aug. 1971

[Follett80]
>   Follett, R.
>   Synthesizing recursive functions with side effects
>   *Artificial Intelligence* **13**, pp. 175-200, 1980

[Gerhart75a]
>   Gerhart, S. L.
>   Correctness-preserving program transformations
>   *Proc. 2nd ACM Symp. on Principles of Programming Languages*, Palo Alto, CA, pp. 54-66, Jan. 1975

[Gerhart75b]
>   Gerhart, S. L.
>   Knowledge about programs: A model and case study
>   *Proc. Intl. Conf. on Reliable Software*, Los Angeles, CA, pp. 83-95, Apr. 1975

[Gerhart75c]

Gerhart, S. L.

Verification operator systems and their application to logical analysis of programs

Colloques IRIA on Proving and Improving Programs, Arc-et-Senans, France, pp. 209-221, July 1975

[GerhartYelowitz76a]

Gerhart, S. L. and Yelowitz, L.

Observations of fallibility in applications of modern programming methodologies

IEEE Trans. Software Engineering **SE-2**(3), pp. 195-207, Sept. 1976

[GerhartYelowitz76b]

Gerhart, S. L. and Yelowitz, L.

Control structure abstractions of the backtracking programming technique

IEEE Trans. Software Engineering **SE-2**(4), pp. 285-292, Dec. 1976

[German74]

German, S. M.

A Program Verifier that Generates Inductive Assertions

Undergraduate thesis, Harvard Univ., Cambridge, MA, May 1974

(Memo TR-19-74, Center for Research in Computing Technology)

[German78]

German, S. M.

Automating proofs of the absence of common runtime errors

Conf. Rec. 5th ACM Symp. on Principles of Programming Languages, Tucson, AZ, pp. 105-118, Jan. 1978

[GermanWegbreit75]

German, S. M. and Wegbreit, B.

A synthesizer of inductive assertions

IEEE Trans. Software Engineering **SE-1**(1), pp. 68-75, Mar. 1975

[Gibb61]
>   Gibb, A.
>   Algorithm 61: Procedures for range arithmetic
>   *Comm. ACM* **4**(7), pp. 319-320, July 1961

[Goldberg74]
>   Goldberg, P. C.
>   Automatic programming
>   *Proc. 4th Informatik Symp.*, Wilbad, West Germany, pp. 347-361, Sept. 1974

[Goto79]
>   Goto, S.
>   Program synthesis from natural deduction proofs
>   *Proc. 6th Intl. Joint Conf. on Artificial Intelligence*, Tokyo, Japan, pp. 339-341, Aug. 1979

[GouldDrongowski74]
>   Gould, J. D. and Drongowski, P.
>   An exploratory study of computer program debugging
>   *Human Factors* **16**, pp. 258-277, 1974

[GrantSackman67]
>   Grant, E. E. and Sackman, H.
>   An exploratory investigation of programmer performance under on-line and off-line conditions
>   *IEEE Trans. on Human Factors in Electronics* **HFE-8**(1), pp. 33-51, March 1967

[Green76]
>   Green, C. C.
>   The design of the PSI program synthesis system
>   *Proc. 2nd Intl. Conf. on Software Engineering*, San Francisco, CA, pp. 4-18, Oct. 1976

[GreifWaldinger74]
>   Greif, I. and Waldinger, R. J.
>   A more mechanical heuristic approach to program verification
>   *Proc. Programming Symp.*, Paris, France, pp. 109-119, Apr. 1974

[Gries74]
Gries, D.
On structured programming - A reply to Smoliar
Comm. ACM **17**(11), pp. 655-657, Nov. 1974

[Gries77]
Gries, D.
Asssignment to subscripted variables
Report TR 77-305, Dept. Computer Science, Cornell Univ., Ithaca, NY, 1977

[Gries81]
Gries, D.
The Science of Programming
Springer-Verlag, New York, NY, 1981

[Hardy75]
Hardy, S.
Synthesis of LISP functions from examples
Adv. Papers 4th Intl. Joint Conf. on Artificial Intelligence, Tbilisi, USSR, pp. 240-245, Sept. 1975

[Harrison77]
Harrison, W. H.
Compiler analysis of the value ranges for variables
IEEE Trans. Software Engineering **SE-3**(3), pp. 243-250, May 1977

[HayesKlahrMostow80]
Hayes-Roth, F., Klahr, P., and Mostow, D. J.
Knowledge acquisition, knowledge refinement, and knowledge programming
Technical Report R-2540-NSF, Rand Corp., May 1980

[Hesse66]
Hesse, M.
Models and Analogies in Science
University of Notre Dame Press, Notre Dame, IN, 1966

[Hikita79]

Hikita, T.

On a class of recursive procedures and equivalent iterative ones
Acta Informatica **12**, pp. 305-320, 1979

[Hoare61]

Hoare, C. A. R.

Algorithm 63: Partition
Comm. ACM **4**(7), p. 321, July 1961

[Hoare69]

Hoare, C. A. R.

An axiomatic basis of computer programming
Comm. ACM **12**(10), pp. 576-580, 583, Oct. 1969

[Hochhauser78]

Hochhauser, S.

Automatic Analysis of Termination and Non-termination
Master's thesis, Weizmann Institute of Science, Rehovot, Israel,
March 1978

[Hogger81]

Hogger, C. J.

Derivation of logic programs
J. ACM **28**(2), pp. 372-392, Apr. 1981

[Howden77]

Howden, W. E.

Symbolic testing and the DISSECT symbolic evaluation system
IEEE Trans. Software Engineering **SE-3**, pp. 266-278, July 1977

[HuetLang78]

Huet, G. and Lang, B.

Proving and applying program transformations expressed with
second-order patterns
Acta Informatica **11**, pp. 31-55, 1978

[IBM70]
International Business Machine,
Scientific subroutine package: Version III programmer's manual
IBM, White Plains, NY, 1970
(Publ. No. 6H20-0205-4)

[JonassenKnuth78]
Jonassen, A. T. and Knuth, D. E.
A trivial algorithm whose analysis isn't
J. Computer and System Sciences **16**(3), pp. 301-322, June 1978

[JouannaudGuiho79]
Jouannaud, J. P. and Guiho, G.
Inference of functions with an interactive system
Machine Intelligence **9**, Wiley, New York, NY, pp. 227-250, 1979

[JouannaudKodratoff80]
Jouannaud, J. P. and Kodratoff, Y.
An automatic construction of LISP programs by transformations of
functions synthesized from their input-output behavior
Intl. J. of Policy Analysis and Information Systems **4**, pp. 331-358,
1980

[Kant81]
Kant, E.
Efficiency in Program Synthesis
UMI Research Press, Ann Arbor, MI, 1981

[Katz76]
Katz, S. M.
Invariants and the Logical Analysis of Programs
Ph.D. Dissertation, Weizmann Institute of Science, Rehovot, Israel,
Sept. 1976

[KatzManna73]
Katz, S. M. and Manna, Z.
A heuristic approach to program verification
Adv. Papers 3rd Intl. Conf. on Artificial Intelligence, Stanford, CA,
pp. 500-512, Aug. 1973

[KatzManna75a]
Katz, S. M. and Manna, Z.
Towards automatic debugging of programs
Proc. Intl. Conf. on Reliable Software, Los Angeles, CA, pp. 143-155,
Apr. 1975

[KatzManna75b]
Katz, S. M. and Manna, Z.
A closer look at termination
Acta Informatica **5**(4), pp. 333-352, Dec. 1975

[KatzManna76]
Katz, S. M. and Manna, Z.
Logical analysis of programs
Comm. ACM **19**(4), pp. 188-206, Apr. 1976

[KieburtzShultis81]
Kieburtz, R. and Shultis, J.
Transformation of FP program schemas
*Proc. Conf. on Functional Programming Languages and Computer
Architecture*, Portsmouth, MA, pp. 41-48, 1981

[King76]
King, J. C.
Symbolic execution and program testing
Comm. ACM **19**(7), pp. 385-391, July 1976

[KirchnerKirchnerJouannaud81]
Kirchner, C., Kirchner, H., and Jouannaud, J. P.
Algebraic manipulations as a unification and matching strategy for
linear equations in signed binary trees
Proc. 7th Intl. Joint Conf. on Artificial Intelligence, Vancouver, Ca-
nada, pp. 1016-1023, Aug. 1981

[Kling71]
Kling, R. E.
*Reasoning by Analogy with Applications to Heuristic Problem Solving:
A Case Study*
Ph.D. Dissertation, Stanford Univ., Stanford, CA, Aug. 1971

[Knuth68]
Knuth, D. E.
The Art of Computer Programming, Volume 1
Addison-Wesley, Reading, MA, 1968

[Knuth74]
Knuth, D. E.
Structured programming with **go to** statements
Computing Surveys **6**(4), pp. 261-301, Dec. 1974

[Kott82]
Kott, L.
Unfold/fold program transformations
Rapport 155, Inst. Nat. de Recherche en Informatique et en Automatique, Le Chesnay, France, Aug. 1982

[Kowalski74]
Kowalski, R. A.
Predicate logic as programming language
Proc. IFIP Cong., Amsterdam, The Netherlands, pp. 569-574, 1974

[LaubschEisenstadt81]
Laubsch, J. and Eisenstadt, M.
Domain specific debugging aids for novice programmers
Proc. 7th Intl. Joint Conf. on Artificial Intelligence, Vancouver, Canada, pp. 964-969, Aug. 1981

[LeeRoeverGerhart79]
Lee, S., deRoever, W. P., and Gerhart, S. L.
The evolution of list-copying algorithms
Proc. 6th ACM Symp. on Principles of Programming Languages, San Antonio, TX, pp. 53-67, Jan. 1979

[London77]
London, R. L.
Perspectives on program verification
In R. T. Yeh, editor, *Current Trends in Programming Methodolgy*, vol. II, pp. 151-172
Prentice-Hall, Englewood Cliffs, NJ, 1977

[Long77]
Long, W. J.
A Program Writer
Ph.D. Dissertation, Mass. Inst. of Technology, Cambridge, MA, Nov. 1977
(Report LCS/TR-187, Lab. for Computer Science)

[Loveman77]
Loveman, D. B.
Program improvement by source-to-source transformation
J. ACM **24**(1), pp. 121-145, Jan. 1977

[LuckhamBuchanan74]
Luckham, D. C. and Buchanan, J. R.
Automatic generation of programs containing conditional statements
Proc. Conf. on Artificial Intelligence and the Simulation of Behavior, Sussex, England, pp. 102-126, July 1974

[LuckhamSuzuki77]
Luckham, D. C. and Suzuki, N.
Proof of termination within a weak logic of programs
Acta Informatica **8**(1), pp. 21-36, Mar. 1977

[LuckhamSuzuki79]
Luckham, D. C. and Suzuki, N.
Verification of array, record, and pointer operations in Pascal
ACM Trans. Programming Languages and Systems **1**(2), pp. 226-244, Oct. 1979

[McCarthy62]
McCarthy, J.
Towards a mathematical science of computation
Proc. Information Processing Cong. 62, North-Holland, Munich, West Germany, pp. 21-28, Aug. 1962

[McDermott79]

McDermott, J.

Learning to use analogies

Proc. 6th Intl. Joint Conf. on Artificial Intelligence, Tokyo, Japan, pp. 568-576, Aug. 1979

[Manna71]

Manna, Z.

Mathematical theory of partial correctness

J. Computer and System Sciences **5**(3), pp. 239-253, June 1971

[MannaNessVuillemin72]

Manna, Z., Ness, S., and Vuillemin, J.

Inductive methods for proving properties of programs

Proc. ACM Conf. on Proving Asserions about Programs, Las Cruces, NM, pp. 27-50, Jan. 1972

[MannaPnueli74]

Manna, Z. and Pnueli, A.

Aximomatic approach to total correctness of programs

Acta Informatica **3**, pp. 243-274, 1974

[MannaWaldinger75]

Manna, Z. and Waldinger, R. J.

Knowledge and reasoning in program synthesis

Artificial Intelligence **6**(2), pp. 175-208, Summer 1975

[MannaWaldinger78]

Manna, Z. and Waldinger, R. J.

The logic of computer programming

IEEE Trans. Software Engineering **SE-4**(3), pp. 199-229, May 1978

[MannaWaldinger79]

Manna, Z. and Waldinger, R. J.

Synthesis: Dreams $\Rightarrow$ programs

IEEE Trans. Software Engineering **SE-5**(4), pp. 294-328, July 1979

[MannaWaldinger80]

Manna, Z. and Waldinger, R. J.

A deductive approach to program synthesis

ACM Trans. Programming Languages and Systems **2**(1), pp. 90-121, Jan. 1980

[MillerGoldstein77]

Miller, M. L. and Goldstein, I. P.

Structured planning and debugging

Proc. 5th Intl. Joint Conf. on Artificial Intelligence, Cambridge, MA, pp. 773-779, Aug. 1977

[Misra77]

Misra, J.

Prospects and limitations of automatic assertion generation for loop programs

SIAM J. Computing **6**(4), pp. 718-729, Dec. 1977

[Misra78]

Misra, J.

An approach to formal definitions and proofs of programming principles

IEEE Trans. Software Engineering **SE-4**(5), pp. 410-413, Sept. 1978

[MooreNewell73]

Moore, J. A. and Newell, A.

How can Merlin understand

In L. Gregg, editor, *Knowledge and Cognition*

Lawrence Erlbaum, 1973

[Moriconi74]

Moriconi, M. S.

Towards the interactive synthesis of assertions

Memo ATP-20, Automatic Theorem Proving Project, Univ. of Texas, Austin, TX, Oct. 1974

[MorrisWegbreit77]
  Morris, J. H. and Wegbreit, B.
  Subgoal induction
  Comm. ACM **20**(4), pp. 209-222, Apr. 1977

[Muralidharan82]
  Muralidharan, M. N.
  A Methodology for Algorithm Development Through Schema Transformations
  Computer Science Program, Indian Inst. of Technology, Kanpur, India, Jan. 1982
  (Ph.D. Dissertation)

[NelsonOppen80]
  Nelson, C. G. and Oppen, D. C.
  Fast decision procedures based on congruence closure
  J. ACM **27**(2), pp. 356-364, 1980

[Netzer76]
  Netzer, I.
  Logical Analysis of Recursive Programs
  Master's thesis, Weizmann Institute of Science, Rehovot, Israel, Apr. 1976

[NewellSimon72]
  Newell, A. and Simon, H. A.
  Human Problem Solving
  Prentice-Hall, Englewood Cliffs, NJ, 1972

[Paige83]
  Paige, R.
  Transformational programming-- Applications to algorithms and systems
  Proc. 10th Symp. on Principles of Programming Languages, Austin, TX, pp. 73-87, Jan. 1983

[PatersonHewitt70]
Paterson, M. S. and Hewitt, C. E.
Comparative schematology
Rec. Project MAC Conf. on Concurrent Systems and Parallel Computation, Woods Hole, MA, pp. 119-128, Dec. 1970

[Phillips77]
Phillips, J. V.
Program inference from traces using multiple knowledge sources
Proc. 5th Intl. Joint Conf. on Artificial Intelligence, Cambridge, MA, p. 812, Aug. 1977

[PhillipsGreen80]
Phillips, J. V. and Green, C. C.
Towards self-described programming environments
Technical Report, Computer Science Dept., Systems Control Inc., Palo Alto, CA, Apr. 1980

[Plaisted80]
Plaisted, D. A.
Abstaction mappings in mechanical theorem proving
Proc. 5th Conf. on Automated Deduction, Les Arcs, France, pp. 264-280, July 1980

[Prywes77]
Prywes, N. S.
Automatic generation of computer programs
Advances in Computers **16**, Academic Press, New York, NY, pp. 57-125, 1977

[RichWaters81]
Rich, C. and Waters, R. C.
Abstraction, inspection and debugging in programming
Memo 634, Artificial Intelligence Lab., Mass. Inst. of Technology, Cambridge, MA, June 1981

[Rohl77]
>Rohl, J. S.
>Converting a class of recursive procedures into non-recursive ones
>*Software - Practice and Experience* **7**(2), pp. 231-238, Mar. 1977

[Ruth76]
>Ruth, G.
>Intelligent program analysis
>*Artificial Intelligence* **7**, pp. 65-85, 1976

[Sacerdoti75]
>Sacerdoti, E. D.
>The nonlinear nature of plans
>*Adv. Papers 4th Intl. Joint Conf. on Artificial Intelligence*, Tbilisi, USSR, pp. 206-214, Sept. 1975

[Sagiv76]
>Sagiv, Y.
>*A study of the automatic debugging of programs*
>Master's thesis, Weizmann Institute of Science, Rehovot, Israel, Aug. 1976

[Sato79]
>Sato, M.
>Towards a mathematical theory of program synthesis
>*Proc. 6th Intl. Joint Conf. on Artificial Intelligence*, Tokyo, Japan, pp. 757-762, Aug. 1979

[Scherlis74]
>Scherlis, W. L.
>*On the Weak Interpretation Method for Extracting Program Properties*
>Undergraduate thesis, Harvard Univ., Cambridge, MA, May 1974

[Scherlis80]
>Scherlis, W. L.
>*Expression Procedures and Program Derivation*
>Ph.D. Dissertation, Stanford Univ., Stanford, CA, Aug. 1980
>(Memo, Dept. Computer Science)

[Schwartz71]
Schwartz, J. T.
An overview of bugs
In R. Rustin, editor, *Debugging Techniques in Large Systems*, pp. 1-16
Prentice-Hall, Englewood Cliffs, NJ, 1971

[Schwartz77]
Schwartz, J. T.
Correct program technology
Report NSO-12, Courant Institute, New York Univ., Sept. 1977

[Shapiro83]
Shapiro, E. Y.
Algorithmic Program Debugging
MIT Press, Cambridge, MA, 1983

[ShawSwartoutGreen75]
Shaw, D., Swartout, W., and Green, C. C.
Infering LISP programs from example problems
Adv. Papers 4th Intl. Joint Conf. on Artificial Intelligence, Tbilisi, USSR, pp. 260-267, Sept. 1975

[Siklossy74]
Siklossy, L.
The synthesis of programs from their properties and the insane heuristic
Proc. 3rd Texas Conf. on Computing Systems, Austin, TX, pp. 5.2.1-5.2.5, 1974

[SiklossySykes75]
Siklossy, L. and Sykes, D. A.
Automatic program synthesis from example problems
Adv. Papers 4th Intl. Joint Conf. on Artificial Intelligence, Tbilisi, USSR, pp. 268-273, Sept. 1975

[Sintzoff72]

Sintzoff, M.

Calculating properties of programs by valuations on specific models

Proc. ACM Conf. on Proving Assertions About Programs **7**(1), Las Cruces, NM, pp. 203-207, Jan. 1972

[Smith80]

Smith, D. R.

A survey of synthesis of LISP programs from examples

Intl. Workshop on Program Construction, Bonas, France, 1980

[Smith82]

Smith, D. R.

Derived preconditions and their use in program synthesis

Proc. 6th Conf. on Automated Deduction, New York, NY, pp. 172-193, June 1982

[Soloway,*etal.*81]

Soloway, E. M., Woolf, B., Rubin, E., and Barth, P.

MENO-II: An intelligent tutoring system for novice programmers

Proc. 7th Intl. Joint Conf. on Artificial Intelligence, Vancouver, Canada, pp. 975-977, Aug. 1981

[Standish73]

Standish, T. A.

Observations and hypotheses about program synthesis mechanisms

Report No. 2780, Bolt, Beranek and Newman, Cambridge, MA, Dec. 1973

[Standish,*etal.*76]

Standish, T. A., Harriman, D. C., Kibler, D. F., and Neighbors, J. M.

The Irvine program transformation catalogue

Report, Dept. Information and Computer Science, Univ. of California, Irvine, CA, Jan. 1976

[StandishKiblerNeighbors76]
Standish, T. A., Kibler, D. F., and Neighbors, J. M.
Improving and refining programs by program manipulation
Proc. ACM Natl. Conf., pp. 509-516, Oct. 1976

[Stefanesco78]
Stefanesco, D. C.
Interactive computer-aided debugging
Report TR-06-78, Center for Research in Computing Technology,
Harvard Univ., Cambridge, MA, 1978

[Steinberg80]
Steinberg, L.
Question ordering in mixed initiative program specification dialogue
Proc. 1st Natl. Conf. on Artificial Intelligence, Stanford, CA, pp. 61-
63, Aug. 1980

[Sternberg77]
Sternberg, R. J.
Intelligence, Information Processing, and Analogical Reasoning
Lawrence Erlbaum, Hillsdale, NJ, 1977

[Summers77]
Summers, P. D.
A methodology for LISP program construction from examples
J. ACM **24**(1), pp. 161-175, Jan. 1977

[Sussman75]
Sussman, G. J.
A Computer Model of Skill Acquisition
American Elsevier, New York, NY, 1975

[SuzukiIshihata77]
Suzuki, N. and Ishihata, K.
Implementation of an array bound checker
*Conf. Rec. 4th ACM Symp. on Principles of Programming
Languages*, Los Angeles, CA, pp. 132-143, Jan. 1977

[Tamir80]

Tamir, M.

ADI: Automatic derivation of invariants

IEEE Trans. Software Engineering **SE-6**(1), pp. 40-48, Jan. 1980

[Teitelman74]

Teitelman, W.

INTERLISP Reference Manual

Xerox Research Center, Palo Alto, CA, 1974

[UlrichMoll77]

Ulrich, J. W. and Moll, R.

Program synthesis by analogy

Proc. ACM Symp. on Artificial Intelligence and Programming Languages, Rochester, NY, pp. 22-28, Aug. 1977

[Villemin81]

Villemin, F. Y.

PAPE: An on-line system for infering procedures from sets of their traces

Proc. 7th Intl. Joint Conf. on Artificial Intelligence, Vancouver, Canada, pp. 1004-1009, Aug. 1981

[Wadler81]

Wadler, P.

Application style programming, program transformations, and list operators

Proc. Conf. on Functional Programming Languages and Computer Architecture, Portsmouth, MA, pp. 25-32, 1981

[Waldinger69]

Waldinger, R. J.

Constructing Programs Automatically Using Theorem Proving

Ph.D. Dissertation, Carnegie-Mellon Univ., Pittsburgh, PA, May 1969

[Waldinger77]
Waldinger, R. J.
Achieving several goals simultaneously
In D. Michie, editor, *Machine Intelligence 8: Machine Representations of Knowledge*, pp. 94-136
Ellis Horwood, Chichester, England, 1977

[WaldingerLevitt74]
Waldinger, R. J. and Levitt, K. N.
Reasoning about programs
Artificial Intelligence 5(3), pp. 235-316, Fall 1974

[WalkerStrong73]
Walker, S. A. and Strong, H. R.
Characterizations of flowchartable recursions
J. Computer and System Sciences 7(4), pp. 404-447, Aug. 1973

[Wand77]
Wand, M.
Continuation-based program transformation strategies
Report 61, Computer Science Dept., Indiana Univ., Bloomington, IN, March 1977

[Warren76]
Warren, D. H. D.
Generating conditional plans and programs
Proc. Conf. on Artificial Intelligence and Simulation on Behavior, Edinburgh, Scotland, pp. 344-354, July 1976

[Waters77]
Waters, R. C.
A method based in plans for understanding how a loop implements a computation
Working paper 150, Artificial Intelligence Lab., Mass. Inst. of Technology, Cambridge, MA, July 1977

[Waters82]

Waters, R. C.

The programmer's apprentice: Knowledge based program editing

IEEE Trans. Software Engineering **SE-8**(1), pp. 1-12, Jan. 1982

[Wegbreit74]

Wegbreit, B.

The synthesis of loop predicates

Comm. ACM **17**(2), pp. 102-112, Feb. 1974

[Wegbreit75]

Wegbreit, B.

Property extraction in well-founded property sets

IEEE Trans. Software Engineering **SE-1**(3), pp. 270-285, Sept. 1975

[Wegbreit76]

Wegbreit, B.

Goal-directed program transformation

IEEE Trans. Software Engineering **SE-2**, pp. 69-80, 1976

[WegbreitSpitzen76]

Wegbreit, B. and Spitzen, J. M.

Proving properties of complex data structures

J. ACM **23**(2), pp. 389-396, Apr. 1976

[Wensley59]

Wensley, J. H.

A class of non-analytical iterative processes

Computer J. **1**(4), pp. 163-167, Jan. 1959

[Wertz79]

Wertz, H.

A system to improve incorrect programs

Proc. 4th Intl. Conf. on Software Engineering, Munich, Germany, pp. 286-293, Sept. 1979

[WeyukerOstrand79]
Weyuker, E. J. and Ostrand, T. J.
Theories of program testing and the application of revealing sub-domains
Report 008, Dept. of Computer Science, New York Univ., New York, NY, Feb. 1979

[Wilber76]
Wilber, B. M.
A QLISP reference manual
Technical note 118, Artificial Intelligence Center, Stanford Research Institute, Menlo Park, CA, Mar. 1976

[Winston80]
Winston, P. H.
Learning and reasoning by analogy
Comm. ACM **23**(12), pp. 689-703, Dec. 1980

[Wirth71]
Wirth, N.
Program development by stepwise refinement
Comm. ACM **14**(4), pp. 221-227, Apr. 1971

[Wirth73]
Wirth, N.
Systematic Programming: An Introduction
Prentice-Hall, Englewood Cliffs, NJ, 1973

[Wirth74]
Wirth, N.
On the composition of well-structured programs
Computing Surveys **6**(4), pp. 247-259, Dec. 1974

[Wirth76]
Wirth, N.
Algorithms + Data Structures = Programs
Prentice-Hall, Englewood Cliffs, NJ, 1976

[Wood80]
　　Wood, R. J.
　　Computer aided program synthesis
　　Report TR-861, Univ. of Maryland, Jan. 1980

[Wossner, *et al.*78]
　　Wossner, H., Pepper, P., Partsch, H., and Bauer, F. L.
　　Special transformation techniques
　　Proc. Intl. Summer School on Program Construction, Marktoberdorf,
　　West Germany, 1978

[YelowitzDuncan77]
　　Yelowitz, L. and Duncan, A. G.
　　Abstractions, instantiations, and proofs of marking algorithms
　　Proc. Conf. on Artificial Intelligence and Programming Languages,
　　Rochester, NY, pp. 13-21, Aug. 1977

[Yohe79]
　　Yohe, J. M.
　　Software for interval arithmetic: A reasonably portable package
　　ACM Trans. on Mathematical Software **5**(1), pp. 50-63, March 1979

Name Index

*In Jerusalem he had machines designed by engineers ... and
his name spread far and wide.*

—*Chronicles*

בריך רחמנא דסעיין

PROGRESS IN COMPUTER SCIENCE
Already published